Programming 8-bit
PIC Microcontrollers in C

with Interactive Hardware Simulation

Programming 8-bit
PIC Microcontrollers in C

with Interactive Hardware Simulation

Martin P. Bates

AMSTERDAM • BOSTON • HEIDELBERG • LONDON
NEW YORK • OXFORD • PARIS • SAN DIEGO
SAN FRANCISCO • SINGAPORE • SYDNEY • TOKYO

Newnes is an imprint of Elsevier

ELSEVIER

Newnes

Newnes is an imprint of Elsevier
30 Corporate Drive, Suite 400, Burlington, MA 01803, USA
Linacre House, Jordan Hill, Oxford OX2 8DP, UK

 Recognizing the importance of preserving what has been written, Elsevier prints its
books on acid-free paper whenever possible.

Library of Congress Cataloging-in-Publication Data
Application submitted.

British Library Cataloguing-in-Publication Data
A catalogue record for this book is available from the British Library.

ISBN: 978-0-7506-8960-1

For information on all Newnes publications
visit our Web site at: www.books.elsevier.com

Transferred to Digital Printing, 2010
Printed and bound in the United Kingdom
Typeset by Charon Tec Ltd., A Macmillan Company. (www.macmillansolutions.com)

Contents

Foreword

Embedded microcontrollers are everywhere today. In the average household you will find them far beyond the obvious places like cell phones, calculators, and MP3 players. Hardly any new appliance arrives in the home without at least one controller and, most likely, there will be several—one microcontroller for the user interface (buttons and display), another to control the motor, and perhaps even an overall system manager. This applies whether the appliance in question is a washing machine, garage door opener, curling iron, or toothbrush. If the product uses a rechargeable battery, modern high density battery chemistries require intelligent chargers.

A decade ago, there were significant barriers to learning how to use microcontrollers. The cheapest programmer was about a hundred dollars and application development required both erasable windowed parts—which cost about ten times the price of the one time programmable (OTP) version—and a UV Eraser to erase the windowed part. Debugging tools were the realm of professionals alone. Now most microcontrollers use Flash-based program memory that is electrically erasable. This means the device can be reprogrammed in the circuit—no UV eraser required and no special packages needed for development. The total cost to get started today is about twenty-five dollars which buys a PICkit™ 2 Starter Kit, providing programming and debugging for many Microchip Technology Inc. MCUs. Microchip Technology has always offered a free Integrated Development Environment (IDE) including an assembler and a simulator. It has never been less expensive to get started with embedded microcontrollers than it is today.

While MPLAB® includes the assembler for free, assembly code is more cumbersome to write, in the first place, and also more difficult to maintain. Developing code using C frees the programmer from the details of multi-byte math and paging and generally improves code readability and maintainability. CCS and Hi-Tech both offer free "student" versions of the compiler to get started and even the full versions are relatively inexpensive once the savings in development time has been taken into account.

While the C language eliminates the need to learn the PIC16 assembly language and frees the user from managing all the details, it is still necessary to understand the architecture. Clocking options, peripherals sets, and pin multiplexing issues still need to be solved. Martin's book guides readers, step-by-step, on the journey from "this is a micro-controller" to "here's how to complete an application." Exercises use the fully featured PIC16F877A, covering the architecture and device configuration. This is a good starting point because other PIC16s are similar in architecture but differ in terms of IO lines, memory, or peripheral sets. An application developed on the PIC16F877A can easily be transferred to a smaller and cheaper midrange PICmicro. The book also introduces the peripherals and shows how they can simplify the firmware by letting the hardware do the work.

MPLAB®, Microchip's Integrated Development Environment, is also covered. MPLAB includes an editor and a simulator and interfaces with many compilers, including the CCS compiler used in this book. Finally, the book includes the Proteus® simulator which allows complete system simulation, saving time and money on prototype PCBs.

Dan Butler
Principal Applications Engineer
Microchip Technology Inc.

Preface

This book is the third in a series, including

- PIC Microcontrollers: An Introduction to Microelectronic Systems.

- Interfacing PIC Microcontrollers: Embedded Design by Interactive Simulation.

- Programming 8-bit PIC Microcontrollers in C: With Interactive Hardware Simulation.

It completes a set that introduces embedded application design using the Microchip PIC® range, from Microchip Technology Inc. of Arizona. This is the most popular microcontroller for education and training, which is also rapidly gaining ground in the industrial and commercial sectors. *Interfacing PIC Microcontrollers* and *Programming PIC Microcontrollers* present sample applications using the leading design and simulation software for microcontroller based circuits, Proteus VSM® from Labcenter Electronics. Demo application files can be downloaded from the author's support Web site (see later for details) and run on-screen so that the operation of each program can be studied in detail.

The purpose of this book is to

- Introduce C programming specifically for microcontrollers in easy steps.

- Demonstrate the use of the Microchip MPLAB IDE for C projects.

- Provide a beginners' guide to the CCS PCM C compiler for 16 series PICs.

- Explain how to use Proteus VSM to test C applications in simulated hardware.

- Describe applications for the Microchip PICDEM mechatronics board.

- Outline the principles of embedded system design and project development.

C is becoming the language of choice for embedded systems, as memory capacity increases in microcontrollers. Microchip supplies the 18 and 24 series chips specifically designed for C programming. However, C can be used in the less complex 16 series PIC, as long as the applications are relatively simple and therefore do not exceed the more limited memory capacity.

The PIC 16F877A microcontroller is used as the reference device in this book, as it contains a full range of peripherals and a reasonable memory capacity. It was also used in the previous work on interfacing, so there is continuity if the book series is taken as a complete course in PIC application development.

Microcontrollers are traditionally programmed in assembly language, each type having its own syntax, which translates directly into machine code. Some students, teachers, and hobbyists may wish to skip a detailed study of assembler coding and go straight to C, which is generally simpler and more powerful. It is therefore timely to produce a text that does not assume detailed knowledge of assembler and introduces C as gently as possible. Although several C programming books for microcontrollers are on the market, many are too advanced for the C beginner and distract the learner with undesirable detail in the early stages.

This text introduces embedded programming techniques using the simplest possible programs, with on-screen, fully interactive circuit simulation to demonstrate a range of basic techniques, which can then be applied to your own projects. The emphasis is on simple working programs for each topic, with hardware block diagrams to clarify system operation, full circuit schematics, simulation screenshots, and source code listings, as well as working downloads of all examples. Students in college courses and design engineers can document their projects to a high standard using these techniques. Each part concludes with a complete set of self-assessment questions and assignments designed to complete the learning package.

An additional feature of this book is the use of Proteus VSM (virtual system modeling). The schematic capture component, ISIS, allows a circuit diagram to be created using an extensive library of active components. The program is attached to the microcontroller, and the animated schematic allows the application to be comprehensively debugged before downloading to hardware. This not only saves time for the professional engineer but provides an excellent learning tool for the student or hobbyist.

Links, Resources, and Acknowledgments

Microchip Technology Inc. (www.microchip.com)

Microchip Technology Inc. is a manufacturer of PIC® microcontrollers and associated products. I gratefully acknowledge the support and assistance of Microchip Inc. in the development of this book and the use of the company trademarks and intellectual property. Special thanks are due to John Roberts of Microchip UK for his assistance and advice. The company Web site contains details of all Microchip hardware, software, and development systems. MPLAB IDE (integrated development system) must be downloaded and installed to develop new applications using the tools described in this book. The data sheet for the PIC 16F877A microcontroller should also be downloaded as a reference source.

PIC, PICmicro, MPLAB, MPASM, PICkit, dsPIC, and PICDEM are trademarks of Microchip Technology Inc.

Labcenter Electronics (www.labcenter.co.uk)

Labcenter Electronics is the developer of Proteus VSM (virtual system modeling), the most advanced cosimulation system for embedded applications. I gratefully acknowledge the assistance of the Labcenter team, especially John Jameson, in the development of this series of books. A student/evaluation version of the simulation software may be downloaded from www.proteuslite.com. A special offer for ISIS Lite, ProSPICE Lite, and the 16F877A simulator model can be found at www.proteuslite.com/register/ipmbundle.htm.

Proteus VSM, ISIS, and ARES are trademarks of Labcenter Electronics Ltd.

Custom Computer Services Inc. (www.ccsinfo.com)

Custom Computer Services Inc. specializes in compilers for PIC microcontrollers. The main range comprises PCB compiler for 12-bit PICs, PCM for 16-bit, and PCH for the 18 series chips. The support provided by James Merriman at CCS Inc. is gratefully acknowledged. The manual for the CCS compiler should be downloaded from the company Web site (Version 4 was used for this book). A 30-day trial version, which will compile code for the 16F877A, is available at the time of writing.

The Author's Web Site (www.picmicros.org.uk)

This book is supported by a dedicated Web site, www.picmicros.org.uk. All the application examples in the book may be downloaded free of charge and tested using an evaluation version of Proteus VSM. The design files are locked so that the hardware configuration cannot be changed without purchasing a suitable VSM license. Similarly, the attached program cannot be modified and recompiled without a suitable compiler license, available from the CCS Web site. Special manufacturer's offers are available via links at my site. This site is hosted by www.larrytech.com and special thanks are due to Gabe Hudson of Larrytech® Internet Services for friendly maintenance and support.

I can be contacted at the e-mail address martin@picmicros.org.uk with any queries or comments related to the PIC book series.

Finally, thanks to Julia for doing the boring domestic stuff so I can do the interesting technical stuff.

About the Author

Martin P. Bates is the author of *PIC Microcontrollers*, Second Edition. He is currently lecturing on electronics and electrical engineering at Hastings College, UK. His interests include microcontroller applications and embedded system design.

Introduction

The book is organized in five parts. Part 1 includes an overview of the PIC microcontroller internal architecture, describing the features of the 16F877A specifically. This chip is often used as representative of the 16 series MCUs because it has a full range of peripheral interfaces. All 16 series chips have a common program execution core, with variation mainly in the size of program and data memory. During programming, certain operational features are configurable: type of clock circuit, watchdog timer enable, reset mechanisms, and so on. Internal features include the file register system, which contains the control registers and RAM block, and a nonvolatile EEPROM block. The parallel ports provide the default I/O for the MCU, but most pins have more than one function. Eight analog inputs and serial interfaces (UART, SPI, and I^2C) are brought out to specific pins. The hardware features of all these are outlined, so that I/O programming can be more readily understood later on. The application development process is described, using only MPLAB IDE in this initial phase. A sample C program is edited, compiled, downloaded, and tested to demonstrate the basic process and the generated file set analyzed. The debugging features of MPLAB are also outlined: run, single step, breakpoints, watch windows, and so on. Disassembly of the object code allows the intermediate assembly language version of the C source program to be analyzed.

Part 2 introduces C programming, using the simplest possible programs. Input and output are dealt with immediately, since this is the key feature of embedded programs. Variables, conditional blocks (IF), looping (WHILE, FOR) are quickly introduced, with a complete example program. Variables and sequence control are considered in a little more detail and functions introduced. This leads on to library functions for operating timers and ports. The keypad and alphanumeric LCD are used in a simple calculator program. More data types (long integers, floating point numbers, arrays, etc.) follow as well as assembler directives and the purpose of the header file. Finally, insertion of assembler into C programs is outlined.

Part 3 focuses on programming input and output operations using the CCS C library functions. These simplify the programming process, with a small set of functions usually providing all the initialization and operating sequences required. Example programs for analog input and the use of interrupts and timers are developed and the serial port functions demonstrated in sample applications. The advantages of each type of serial bus are compared, and examples showing the connection of external serial EEPROM for data storage and a digital to analog converter output are provided. These applications can be tested in VSM, but this is not essential; use of VSM is optional throughout the book.

Part 4 focuses specifically on the PICDEM mechatronics board from Microchip. This has been selected as the main demonstration application, as it is relatively inexpensive and contains a range of features that allow the features of a typical mechatronics system to be examined: input sensors (temperature, light, and position) and output actuators (DC and stepper motor). These are tested individually then the requirements of a temperature controller outlined. Operation of the 3.5-digit seven-segment LCD is explained in detail, as this is not covered elsewhere. A simulation version of the board is provided to aid further application design and implementation.

Part 5 outlines some principles of software and hardware design and provides some further examples. A simple temperature controller provides an alternative design to that based on the mechatronics board, and a data logger design is based on another standard hardware system, which can be adapted to a range of applications—the BASE board. Again, a full-simulation version is provided for testing and further development work. This is followed by a section on operating systems, which compares three program design options: a polling loop, interrupt driven systems, and real-time operating systems. Consideration of criteria for the final selection of the MCU for a given application and some general design points follow.

Three appendices (A, B, and C) cover hardware design using ISIS schematic capture, software design using CCS C, and system testing using Proteus VSM. These topics are separated from the main body of the book as they are related more to specific products. Taken together, MPLAB, CCS C, and Proteus VSM constitute a complete learning/design package, but using them effectively requires careful study of product-specific tutorials. VSM, in particular, has comprehensive, well-designed help files; and it is therefore unnecessary to duplicate that material here. Furthermore, as with all good design tools, VSM evolves very quickly, so a detailed tutorial quickly becomes outdated.

Appendix D compares alternative compilers, and application development areas are identified that would suit each one. Appendix E provides a summary of CCS C syntax requirements, and Appendix F contains a list of the CCS C library functions provided with the compiler, organized in functional groups for ease of reference. These are intended to provide a convenient reference source when developing CCS C programs, in addition to the full CCS compiler reference manual.

Each part of the book is designed to be as self-contained as possible, so that parts can be skipped or studied in detail, depending on the reader's previous knowledge and interests. On the other hand, the entire book should provide a coherent narrative leading to a solid grounding in C programming for embedded systems in general.

requirements, and Appendix F contains a list of the CCS C library functions provided with the compiler, organized in functional groups for ease of reference. These are intended to provide a convenient reference source when developing CCS C programs, in addition to the full CCS compiler reference manual.

Each part of the book is designed to be as self-contained as possible, so that parts can be skipped or studied in detail depending on the reader's previous knowledge and interests. On the other hand, the entire book should provide a coherent narrative leading to a solid grounding in C programming for embedded systems in general.

PART 1

PIC Microcontroller Systems

1.1 PIC16 Microcontrollers

- MCU features
- Program execution
- RAM file registers
- Other PIC chips

The microcontroller unit (MCU) is now big, or rather small, in electronics. It is one of the most significant developments in the continuing miniaturization of electronic hardware. Now, even trivial products, such as a musical birthday card or electronic price tag, can include an MCU. They are an important factor in the digitization of analog systems, such as sound systems or television. In addition, they provide an essential component of larger systems, such as automobiles, robots, and industrial systems. There is no escape from microcontrollers, so it is pretty useful to know how they work.

The computer or digital controller has three main elements: input and output devices, which communicate with the outside world; a processor, to make calculations and handle data operations; and memory, to store programs and data. Figure 1.1 shows these in a little more detail. Unlike the conventional microprocessor system (such as a PC), which has separate chips on a printed circuit board, the microcontroller contains all these elements in one chip. The MCU is essentially a computer on a chip; however, it still needs input and output devices, such as a keypad and display, to form a working system.

The microcontroller stores its program in ROM (read only memory). In the past, UV (ultraviolet) erasable programmable ROM (EPROM) was used for prototyping or

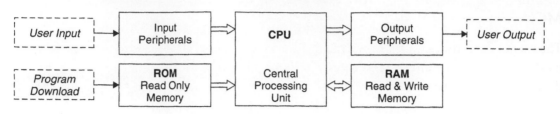

Figure 1.1: Elements of a Digital Controller

small batch production, and one-time programmable ROM for longer product runs. Programmable ROM chips are programmed in the final stages of manufacture, while EPROM could be programmed by the user.

Flash ROM is now normally used for prototyping and low-volume production. This can be programmed in circuit by the user after the circuit has been built. The prototyping cycle is faster, and software variations are easier to accommodate. We are all now familiar with flash ROM as used in USB memory sticks, digital camera memory, and so on, with Gb (10^9 byte) capacities commonplace.

The range of microcontrollers available is expanding rapidly. The first to be widely used, the Intel 8051, was developed alongside the early Intel PC processors, such as the 8086. This device dominated the field for some time; others emerged only slowly, mainly in the form of complex processors for applications such as engine management systems. These devices were relatively expensive, so they were justified only in high-value products. The potential of microcontrollers seems to have been realized only slowly.

The development of flash ROM helped open up the market, and Microchip was among the first to take advantage. The cheap and reprogrammable PIC16F84 became the most widely known, rapidly becoming the number one device for students and hobbyists. On the back of this success, the Microchip product range rapidly developed and diversified. The supporting development system, MPLAB, was distributed free, which helped the PIC to dominate the low-end market.

Flash ROM is one of the technical developments that made learning about microsystems easier and more interesting. Interactive circuit design software is another. The whole design process is now much more transparent, so that working systems are more quickly achievable by the beginner. Low-cost in-circuit debugging is another technique that helps get the final hardware up and running quickly, with only a modest expenditure on development tools.

MCU Features

The range of microcontrollers now available developed because the features of the MCU used in any particular circuit must be as closely matched as possible to the actual needs of the application. Some of the main features to consider are

- Number of inputs and outputs.

- Program memory size.

- Data RAM size.

- Nonvolatile data memory.

- Maximum clock speed.

- Range of interfaces.

- Development system support.

- Cost and availability.

The PIC16F877A is useful as a reference device because it has a minimal instruction set but a full range of peripheral features. The general approach to microcontroller application design followed here is to develop a design using a chip that has spare capacity, then later select a related device that has the set of features most closely matching the application requirements. If necessary, we can drop down to a lower range (PIC10/12 series), or if it becomes clear that more power is needed, we can move up to a higher specification chip (PIC18/24 series). This is possible as all devices have the same core architecture and compatible instructions sets.

The most significant variation among PIC chips is the instruction size, which can be 12, 14, or 16 bits. The A suffix indicates that the chip has a maximum clock speed of 20 MHz, the main upgrade from the original 16F877 device. These chips can otherwise be regarded as identical, the suffix being optional for most purposes. The 16F877A pin-out is seen in Figure 1.2 and the internal architecture in Figure 1.3. The latter is a somewhat simplified version of the definitive block diagram in the data sheet.

Program Execution

The chip has 8 k (8096 × 14 bits) of flash ROM program memory, which has to be programmed via the serial programming pins PGM, PGC, and PGD. The fixed-length

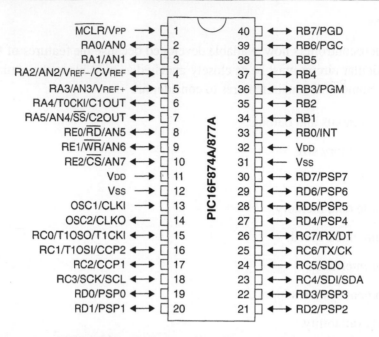

```
MCLR/Vpp         →  [ 1        40 ]  ←→ RB7/PGD
RA0/AN0          ←→ [ 2        39 ]  ←→ RB6/PGC
RA1/AN1          ←→ [ 3        38 ]  ←→ RB5
RA2/AN2/VREF-/CVREF ←→ [ 4     37 ]  ←→ RB4
RA3/AN3/VREF+    ←→ [ 5        36 ]  ←→ RB3/PGM
RA4/T0CKI/C1OUT  ←→ [ 6        35 ]  ←→ RB2
RA5/AN4/SS/C2OUT ←→ [ 7        34 ]  ←→ RB1
RE0/RD/AN5       ←→ [ 8        33 ]  ←→ RB0/INT
RE1/WR/AN6       ←→ [ 9        32 ]  ←  VDD
RE2/CS/AN7       ←→ [ 10       31 ]  ←  VSS
VDD              →  [ 11       30 ]  ←→ RD7/PSP7
VSS              →  [ 12       29 ]  ←→ RD6/PSP6
OSC1/CLKI        →  [ 13       28 ]  ←→ RD5/PSP5
OSC2/CLKO        ←  [ 14       27 ]  ←→ RD4/PSP4
RC0/T1OSO/T1CKI  ←→ [ 15       26 ]  ←→ RC7/RX/DT
RC1/T1OSI/CCP2   ←→ [ 16       25 ]  ←→ RC6/TX/CK
RC2/CCP1         ←→ [ 17       24 ]  ←→ RC5/SDO
RC3/SCK/SCL      ←→ [ 18       23 ]  ←→ RC4/SDI/SDA
RD0/PSP0         ←→ [ 19       22 ]  ←→ RD3/PSP3
RD1/PSP1         ←→ [ 20       21 ]  ←→ RD2/PSP2
```

PIC16F874A/877A

Figure 1.2: 16F877 Pin-out (reproduced by permission of Microchip Inc.)

instructions contain both the operation code and operand (immediate data, register address, or jump address). The mid-range PIC has a limited number of instructions (35) and is therefore classified as a RISC (reduced instruction set computer) processor.

Looking at the internal architecture, we can identify the blocks involved in program execution. The program memory ROM contains the machine code, in locations numbered from 0000h to 1FFFh (8 k). The program counter holds the address of the current instruction and is incremented or modified after each step. On reset or power up, it is reset to zero and the first instruction at address 0000 is loaded into the instruction register, decoded, and executed. The program then proceeds in sequence, operating on the contents of the file registers (000–1FFh), executing data movement instructions to transfer data between ports and file registers or arithmetic and logic instructions to process it. The CPU has one main working register (W), through which all the data must pass.

If a branch instruction (conditional jump) is decoded, a bit test is carried out; and if the result is true, the destination address included in the instruction is loaded into the program counter to force the jump. If the result is false, the execution sequence continues unchanged. In assembly language, when CALL and RETURN are used to implement

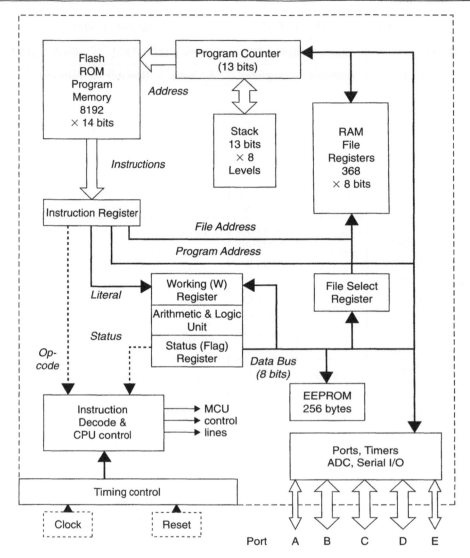

Figure 1.3: PIC16F877 MCU Block Diagram

subroutines, a similar process occurs. The stack is used to store return addresses, so that the program can return automatically to the original program position. However, this mechanism is not used by the CCS C compiler, as it limits the number of levels of subroutine (or C functions) to eight, which is the depth of the stack. Instead, a simple GOTO instruction is used for function calls and returns, with the return address computed by the compiler.

Table 1.2: PIC Microcontroller Types

MCU	Pins	Data Word (bits)	Program Memory (bytes)	Typical Instruction Set	Speed MIPS	Description
10FXXX	= 6	8	≤512	33 × 12 bits	≤2	Low pin count, small form factor, cheap, no EEPROM, no low-power, assembler program
12FXXX	= 8	8	≤2 kB	12/14 bits	≤0.5	Low pin count, small form factor, cheap, EEPROM, 10-bit ADC, some low power, assembler
16FXXX	≤64	8	≤14 kB	35 × 14 bits	≤5	Mid-range, UART, I2C, SPI, many low power, C or assembler program
18FXXXX	≤100	8	≤128 kB	75 × 16 bits	≤16	High range, CAN, USB J series 3V supply, C program
24FXXXX	≤100	16	≤128 kB	76 × 24 bits	= 16	Power range, 3V supply, no EEPROM, data RAM ≤8 kB, C program

available in the low pin count (LPC) ranges (10/12 series), while the power ranges are expanding rapidly. In addition are those listed in the 24HXXXX range, which runs at 40 MIPS, and the dsPIC (digital signal processor) high-specification range.

1.2   PIC16 MCU Configuration

- Clock oscillator types

- Watchdog, power-up, brown-out timers

- Low-voltage programming

- Code protection

- In-circuit debug mode

When programming the PIC microcontroller, certain operational modes must be set prior to the main program download. These are controlled by individual bits in a special configuration register separated from the main memory block. The main options are as follows.

Clock Options

The '877 chip has two main clock modes, CR and XT. The CR mode needs a simple capacitor and resistor circuit attached to CLKIN, whose time constant (C × R) determines the clock period. R should be between 3 k and 100k, and C greater than 20 pF. For example, if R = 10 kΩ and C = 10 nF, the clock period will be around 2 × C × R = 200 μs (calculated from the CR rise/fall time) and the frequency about 5 kHz. This option is acceptable when the program timing is not critical.

The XT mode is the one most commonly used, since the extra component cost is small compared with the cost of the chip itself and accurate timing is often a necessity. An external crystal and two capacitors are fitted to CLKIN and CLKOUT pins. The crystal frequency in this mode can be from 200 kHz to 4 MHz and is typically accurate to better than 50 ppm (parts per million) or 0.005%. A convenient value is 4 Mz, as this is the maximum frequency possible with a standard crystal and gives an instruction execution time of 1.000 μs (1 million instructions per second, or 1 Mip).

A low-speed crystal can be used to reduce power consumption, which is proportional to clock speed in CMOS devices. The LP (low-power) mode supports the clock frequency range 32–200 kHz. To achieve the maximum clock speed of 20 MHz, a high-speed (HS) crystal is needed, with a corresponding increase in power consumption.

The MCU configuration fuses must be set to the required clock mode when the chip is programmed. Many PIC chips now have an internal oscillator, which needs no external components. It is more accurate than the RC clock but less accurate than a crystal. It typically runs at 8 MHz and can be calibrated in the chip configuration phase to provide a more accurate timing source.

Configuration Options

Apart from the clock options, several other hardware options must be selected.

Watchdog Timer

When enabled, the watchdog timer (WDT) automatically resets the processor after a given period (default 18 ms). This allows, for example, an application to escape from an endless loop caused by a program bug or run-time condition not anticipated by the software designer. To maintain normal operation, the WDT must be disabled or reset within the program loop before the set time-out period has expired. It is therefore important to set the MCU configuration bits to disable the WDT if it is not intended to use this feature. Otherwise, the program is liable to misbehave, due to random resetting of the MCU.

Power-up Timer

The power-up timer (PuT) provides a nominal 72 ms delay between the power supply voltage reaching the operating value and the start of program execution. This ensures that the supply voltage is stable before the clock starts up. It is recommended that it be enabled as a precaution, as there is no adverse effect on normal program execution.

Oscillator Start-up Timer

After the power-up timer has expired, a further delay allows the clock to stabilize before program execution begins. When one of the crystal clock modes is selected, the CPU waits 1024 cycles before the CPU is enabled.

Brown-out Reset (BoR)

It is possible for a transitory supply voltage drop, or brown-out, to disrupt the MCU program execution. When enabled, the brown-out detection circuit holds the MCU in reset while the supply voltage is below a given threshold and releases it when the supply has recovered. In CCS C, a low-voltage detect function triggers an interrupt that allows the program to be restarted in an orderly way.

Code Protection (CP)

The chip can be configured during programming to prevent the machine code being read back from the chip to protect commercially valuable or secure code. Optionally, only selected portions of the program code may be write protected (see WRT_X% later).

In-Circuit Programming and Debugging

Most PIC chips now support in-circuit programming and debugging (ICPD), which allows the program code to be downloaded and tested in the target hardware, under the control of the host system. This provides a final test stage after software simulation has been used to eliminate most of the program bugs. MPLAB allows the same interface to be used for debugging in both the simulation and in-circuit modes. The slight disadvantage of this option is that care must be taken that any application circuit connected to the programming/ICPD pins does not interfere with the operation of these features. It is preferable to leave these pins for the exclusive use of the ICPD system. In addition, a small section of program memory is required to run the debugging code.

Low-Voltage Programming Mode

The low-voltage programming mode can be selected during programming so that the customary high (12V) programming voltage is not needed, and the chip can be programmed at V_{dd} (+5V). The downside is that the programming pin cannot then be used for digital I/O. In any case, it is recommended here that the programming pins not be used for I/O by the inexperienced designer, as hardware contention could occur.

Electrically Erasable Programmable Read Only Memory

Many PIC MCUs have a block of nonvolatile user memory where data can be stored during power-down. These data could, for example, be the secure code for an electronic lock or smart card reader. The electrically erasable programmable read only memory (EEPROM) can be rewritten by individual location, unlike flash program ROM. The '877 has a block of 256 bytes, which is a fairly typical value. There is a special read/write sequence to prevent accidental overwriting of the data.

Configuration in C

The preprocessor directive #fuses is used to set the configuration fuses in C programs for PICs. A typical statement is

```
#fuses XT,PUT,NOWDT,NOPROTECT,NOBROWNOUT
```

The options defined in the standard CCS C 16F877 header file are

```
Clock Type Select               LP, XT, HS, RC
Watchdog Timer Enable           WDT, NOWDT
Power Up Timer Enable           PUT, NOPUT
Program Code Protect            PROTECT, NOPROTECT
In Circuit Debugging Enable     DEBUG, NODEBUG
Brownout Reset Enable           BROWNOUT, NOBROWNOUT
Low Voltage Program Enable      LVP, NOLVP
EEPROM Write Protect            CPD, NOCPD
Program Memory Write Protect    WRT_50%, WRT_25%,
(with percentage protected)     WRT_5%, NOWRT
```

The default condition for the fuses if no such directive is included is equivalent to

```
#fuses RC,WDT,NOPUT,BROWNOUT,LVP,NOCPD,NOWRT
```

This corresponds to all the bits of configuration register being default high.

1.3   PIC16 MCU Peripherals

- Digital I/O

- Timers

- A/D converter

- Comparator

- Parallel slave port

- Interrupts

Basic digital input and output (I/O) in the microcontroller uses a bidirectional port pin. The default pin configuration is generally digital input, as this is the safest option if some error has been made in the external connections. To set the pin as output, the corresponding data direction bit must be cleared in the port data direction register (e.g., TRISD). Note, however, that pins connected to the analog-to-digital (A/D) converter default to the analog input mode.

The basic digital I/O hardware is illustrated in simplified form in Figure 1.4, with provision for analog input. The 16 series reference manual shows equivalent circuits for individual pins in more detail. For input, the current driver output is disabled by loading the data direction bit with a 1, which switches off the tristate gate. Data are read into the input data latch from the outside world when its control line is pulsed by the CPU in the course of a port register read instruction. The data are then copied to the CPU working register for processing.

When the port is set up for output, a 0 is loaded into the data direction bit, enabling the current output. The output data are loaded into the data latch from the CPU. A data 1 at the output allows the current driver to source up to 25 mA at 5 V, or whatever the supply voltage is (2–6 V). A data 0 allows the pin to sink a similar current at 0 V.

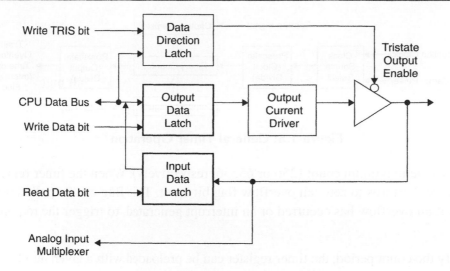

Figure 1.4: I/O Pin Operation

The 16F877 has the following digital I/O ports available:

Port A	RA0–RA5	6 bits
Port B	RB0–RB7	8 bits
Port C	RC0–RC7	8 bits
Port D	RD0–RD7	8 bits
Port E	RE0–RE2	3 bits
Total digital I/O available		33 pins

Most of the pins have alternate functions, which are described later.

Timers

Most microcontrollers provide hardware binary counters that allow a time interval measurement or count to be carried out separately from program execution. For example, a fixed period output pulse train can be generated while the program continues with another task. The features of the timers found in the typical PIC chip are represented in Figure 1.5, but none of those in the '877 has all the features shown.

The count register most commonly is operated by driving it from the internal instruction clock to form a timer. This signal runs at one quarter of the clock frequency; that is, one instruction takes four cycles to execute. Therefore, with a 4-MHz clock, the timer counts in microseconds (1-MHz instruction clock). The number of bits in the timer (8 or 16)

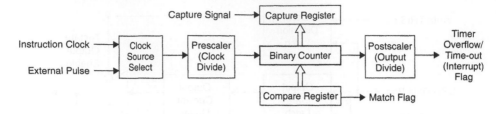

Figure 1.5: General Timer Operation

determines the maximum count (256 or 65536, respectively). When the timer register overflows and returns to zero, an overflow flag bit is set. This flag can be polled (tested) to check if an overflow has occurred or an interrupt generated, to trigger the required action.

To modify the count period, the timer register can be preloaded with a given number. For example, if an 8-bit register is preloaded with the value 156, a time-out occurs after $256 - 156 = 100$ clocks. Many timer modules allow automatic preloading each time it is restarted, in which case the required value is stored in a preload register during timer initialization.

A prescaler typically allows the timer input frequency to be divided by 2, 4, 8, 16, 32, 64, or 128. This extends the maximum count proportionately but at the expense of timer precision. For example, the 8-bit timer driven at 1 MHz with a prescale value of 4 counts up to $256 \times 4 = 1024$ μs, at 4 μs per bit. A postscaler has a similar effect, connected at the output of the counter.

In the compare mode, a separate period register stores a value that is compared with the current count after each clock and the status flag set when they match. This is a more elegant method of modifying the time-out period, which can be used in generating a pulse width modulated (PWM) output. A typical application is to control the output power to a current load, such as a small DC motor—more on this later. In the capture mode, the timer count is captured (copied to another register) at the point in time when an external signal changes at one of the MCU pins. This can be used to measure the length of an input pulse or the period of a waveform.

The '877 has three counter/timer registers. Timer0 has an 8-bit counter and 8-bit prescaler. It can be clocked from the instruction clock or an external signal applied to RA4. The prescaler can also be used to extend the watchdog timer interval (see later), in which case it is not available for use with Timer0. Timer1 has a 16-bit counter and prescaler and can be clocked internally or externally as per Timer0. It offers capture and

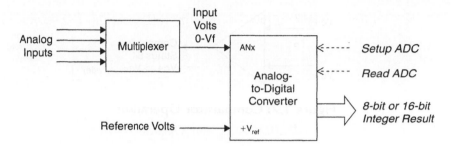

Figure 1.6: ADC Operation

compare modes of operation. Timer2 is another 8-bit counter but has both a prescaler and postscaler (up to 1:16) and a compare register for period control.

Further details are provided in *Interfacing PIC Microcontrollers* by the author and the MCU data books. When programming in C, only a limited knowledge of timer operation is necessary, as the C functions generally take care of the details.

A/D Converter

Certain PIC pins can be set up as inputs to an analog-to-digital converter (ADC). The '877 has eight analog inputs, which are connected to Port A and Port E. When used in this mode, they are referred to as AD0–AD7. The necessary control registers are initialized in CCS C using a set of functions that allow the ADC operating mode and inputs to be selected. An additional "device" directive at the top of the program sets the ADC resolution. An analog voltage presented at the input is then converted to binary and the value assigned to an integer variable when the function to read the ADC is invoked.

The default input range is set by the supply (nominally 0–5 V). If a battery supply is used (which drops over time) or additional accuracy is needed, a separate reference voltage can be fed in at AN2 ($+V_{ref}$) and optionally AN3 ($-V_{ref}$). If only $+V_{ref}$ is used, the lower limit remains 0 V, while the upper is set by the reference voltage. This is typically supplied using a zener diode and voltage divider. The 2.56 V derived from a 2V7 zener gives a conversion factor of 10 mV per bit for an 8-bit conversion. For a 10-bit input, a reference of 4.096 V might be convenient, giving a resolution of 4 mV per bit. The essentials of ADC operation are illustrated in Figure 1.6.

Comparator

The comparator (Figure 1.7) is an alternative type of analog input found in some microcontrollers, such as the 16F917 used in the mechatronics board described later.

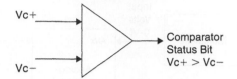

Figure 1.7: Comparator Operation

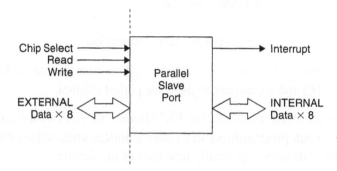

Figure 1.8: Parallel Slave Port Operation

It compares the voltage at a pair of inputs, and a status bit is set if the C+ pin is higher than C−. The comparator status bit may also be monitored at an output pin. The '917 has two such comparator modules; they are enabled using a system function to set the operating mode. The '877 has no comparators, so the ADC must be used instead.

Parallel Slave Port

The parallel slave port on the '877 chip is designed to allow parallel communications with an external 8-bit system data bus or peripheral (Figure 1.8). Port D provides the eight I/O data pins, and Port E three control lines: Read, Write, and Chip Select. If data are to be input to the port, the pin data direction is set accordingly and data presented to Port D. The chip select input must be set low and the data latched into the port data register by taking the write line low. Conversely, data can be read from the port using the read line. Either operation can initiate an interrupt.

Interrupts

Interrupts can be generated by various internal or external hardware events. They are studied in more detail later in relation to programming peripheral operations. However, at this stage, it is useful to have some idea about the interrupt options provided within the MCU. Table 1.3 lists the devices that can be set up to generate an interrupt.

Table 1.3: Interrupts Sources in the PIC16F877

Interrupt Source	Interrupt Trigger Event	Interrupt Label
Timers		
Timer0	Timer0 register overflow	`INT_TIMER0`
Timer1	Timer1 register overflow	`INT_TIMER1`
CCP1	Timer1 capture or compare detected	`INT_CCP1`
Timer2	Timer2 register overflow	`INT_TIMER2`
CCP2	Timer2 capture or compare detected	`INT_CCP2`
Ports		
RB0/INT pin	Change on single pin RB0	`INT_EXT`
Port B pins	Change on any of four pins, RB4–RB7	`INT_RB`
Parallel Slave Port	Data received at PSP (write input active)	`INT_PSP`
Analog Converter	A/D conversion completed	`INT_AD`
Analog Comparator	Voltage compare true	`INT_COMP`
Serial		
UART Serial Port	Received data available	`INT_RDA`
UART Serial Port	Transmit data buffer empty	`INT_TBE`
SPI Serial Port	Data transfer completed (read or write)	`INT_SSP`
I^{2}C Serial Port	Interface activity detected	`INT_SSP`
I^{2}C Serial Port	Bus collision detected	`INT_BUSCOL`
Memory		
EEPROM	Nonvolatile data memory write complete	`INT_EEPROM`

The most effective way of integrating timer operations into an application program is by using a timer interrupt. Figure 1.9 shows a program sequence where a timer is run to generate an output pulse interval. An interrupt routine (ISR) has been written and assigned to the timer interrupt. The timer is set up during program initialization and started by preloading or clearing it. The main program and timer count then proceed concurrently, until a time-out occurs and the interrupt is generated. The main program is suspended and the ISR executed. When finished, the main program is resumed at the original point. If the ISR contains a statement to toggle an output bit, a square wave could be obtained with a period of twice the timer delay.

When interrupts are used in assembly language programs, it is easier to predict the effect, as the programmer has more direct control over the exact sequence of the ISR.

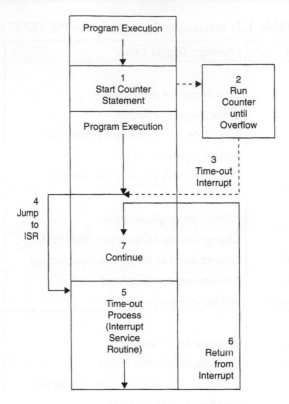

Figure 1.9: Timer Interrupt Process

A C program is generated automatically by the compiler, so the precise timing that results from an interrupt is less obvious. For this reason, the use of a real-time operating system (RTOS) is sometimes preferred in the C environment, especially when programs become more complex. In fact, C was originally developed for precisely this purpose, to write operating systems for computers. C interrupts are considered further in Section 3.2, and RTOS principles are outlined in Section 5.4.

1.4   PIC16 Serial Interfaces

- USART asynchronous link
- SPI synchronous bus
- I2C synchronous bus

Serial data connections are useful because only one or two signal wires are needed, compared with at least eight data lines for a parallel bus plus control signals. The typical

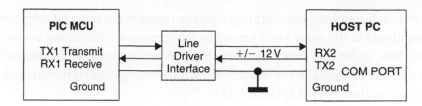

Figure 1.10: USART Operation

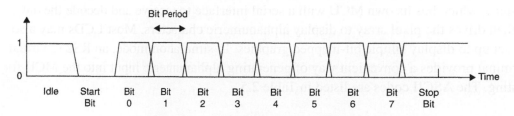

Figure 1.11: USART RS232 Signal

PIC microcontroller offers a choice of serial interfaces. The best one for any given communication channel depends on the distance between nodes, the speed, and the number of hardware connections required.

USART

The universal synchronous/asynchronous receive transmit (USART) device is typically used in asynchronous mode to implement off-board, one-to-one connections. The term *asynchronous* means no separate clock signal is needed to time the data reception, so only a data send, data receive, and ground wires are needed. It is quick and simple to implement if a limited data bandwidth is acceptable.

A common application is connecting the PIC chip to a host PC for uploading data acquired by the MCU subsystem (Figure 1.10). The USART link can send data up to 100 meters by converting the signal to higher-voltage levels (typically ±12V). The digital signal is inverted and shifted to become bipolar (symmetrical about 0V, line negative when inactive) for transmission.

The PIC 16F877 has a dedicated hardware RS232 port, but CCS C allows any pin to be set up as an RS232 port, providing functions to generate the signals in software. The basic form of the signal has 8 data bits and a stop and start bit. The bit period is set by the baud rate. A typical value is 9600 baud, which is about 10k bits per second. The bit period is then about 100 μs, about 1 byte per millisecond, or 1 K byte per second.

The data are transferred between shift registers operating at the same bit rate; the receiver has to be initialized to the same baud setting as the transmitter. Assuming we are looking at TTL level data, in the idle state, the line is high. When it goes low, the receiver clock is started, the data are sampled in the middle of each following data bit period, and data are shifted into the receive register (Figure 1.11).

RS232 is used to access the standard serial LCD display, in which case, line drivers are not necessarily required. ASCII characters and control codes are sent to operate the display, which has its own MCU with a serial interface to receive and decode the data. It then drives the pixel array to display alphanumeric characters. Most LCDs may also be set up to display simple bit-mapped graphics. In simulation mode, an RS232 virtual terminal provides a convenient way of generating alphanumeric input into the MCU for testing. The ASCII codes are listed in Table 2.5.

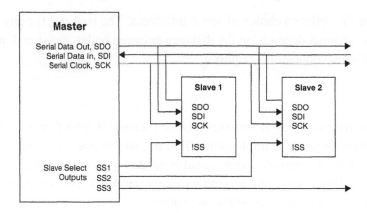

Figure 1.12: SPI Connections

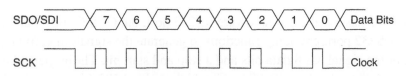

Figure 1.13: SPI Signals

SPI Bus

The serial peripheral interface (SPI) bus provides high-speed synchronous data exchange over relatively short distances (typically within a set of connected boards), using a master/slave system with hardware slave selection (Figure 1.12). One processor must act as a master, generating the clock. Others act as slaves, using the master clock for timing the data send and receive. The slaves can be other microcontrollers or peripherals with an SPI interface. The SPI signals are

- Serial Clock (SCK)

- Serial Data In (SDI)

- Serial Data Out (SDO)

- Slave Select (!SS)

To transfer data, the master selects a slave device to talk to, by taking its SS line low. Eight data bits are then clocked in or out of the slave SPI shift register to or from the master (Figure 1.13). No start and stop bits are necessary, and it is much faster than RS232. The clock signal runs at the same speed as the master instruction clock, that is, 5 MHz when the chip is running at the maximum 20 MHz (16 series MCUs).

I^2C Bus

The interintegrated circuit (I^2C) bus is designed for short-range communication between chips in the same system using a software addressing system. It requires only two signal wires and operates like a simplified local area network. The basic form of the hardware and data signal are illustrated in Figures 1.14 and 1.15.

The I^2C slave chips are attached to a two-wire bus, which is pulled up to logic 1 when idle. Passive slave devices have their register or location addresses determined by a combination of external input address code pins and fixed internal decoding. If several memory devices are connected to the bus, they can be mapped into a continuous address space. The master sends data in 8-bit blocks, with a synchronous clock pulse alongside each bit. As for SPI, the clock is derived from the instruction clock, up to 5 MHz at the maximum clock rate of 20 MHz.

To send a data byte, the master first sends a control code to set up the transfer, then the 8-bit or 10-bit address code, and finally the data. Each byte has a start and acknowledge bit, and each byte must be acknowledged before the next is sent, to improve reliability.

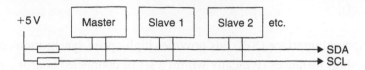

Figure 1.14: I²C Connections

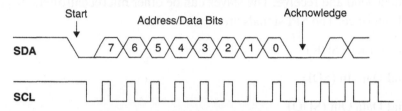

Figure 1.15: I²C Signals

The sequence to read a single byte requires a total of 5 bytes to complete the process, 3 to set the address, and 2 to return the data. Thus, a substantial software overhead is involved. To alleviate this problem, data can be transferred in continuous blocks (memory page read/write), which speeds up the transmission.

1.5   PIC16 MPLAB Projects

- MPLAB C Project

- Project Files

The PIC microcontroller program comprises a list of machine code instructions, decoded and executed in sequence, resulting in data movement between registers, and arithmetic and logic operations. MCU reset starts execution at address zero, and the instructions are executed in address order until a program branch is decoded, at which point a new target address is derived from the instruction. A decision is made to take the branch or continue in sequence based on the result of a bit condition test. This process is described in detail in *PIC Microcontrollers* by the author.

The program could be written in raw binary code, but this would require manual interpretation of the instruction set. Therefore, the machine code is generated from assembly code, where each instruction has a corresponding mnemonic form that is more easily recognizable, such as MOVF05,W (move the data at Port A to the working register). This low-level language is fine for relatively simple programs but becomes time consuming for more complex programs. In addition, assembly language is specific to a particular type of processor and, therefore, not "portable." Another level of abstraction is needed, requiring a high-level language.

C has become the universal language for microcontrollers. It allows the MCU memory and peripherals to be controlled directly, while simplifying peripheral setup, calculations, and other program functions. All computer languages need an agreed set of programming language rules. The definitive C reference is *The C Programming Language* by Kernighan and Ritchie, second edition, incorporating ANSI C standards, published in 1983.

A processor-specific compiler converts the standard syntax into the machine code for a particular processor. The compiler package may also provide a set of function libraries, which implement the most commonly needed operations. There is variation between compilers in the library function syntax, but the general rules are the same.

Usually, a choice of compilers is available for any given MCU family. Options for the PIC at time of writing are Microchip's own C18 compiler, Hi-Tech PICC, and CCS C. CCS was selected for the current work because it is specifically designed for the PIC MCU, supports the 16 series devices, and has a comprehensive set of peripheral driver functions.

MPLAB C Project

The primary function of the compiler is to take a source text file PROJNAME.C and convert it to machine code, PROJNAME.HEX. The hex file can then be downloaded to the PIC MCU. The source file must be written in the correct form, observing the conventions of both ANSI C and the specific compiler dialect. The first program we see later in the tutorial section is shown in Listing 1.1.

This can be typed into any text editor, but we normally use the editor in MPLAB, the standard Microchip development system software package. This provides file management, compiler interface and debugging facilities for PIC projects, and may be downloaded free of charge from www.microchip.com. Before starting work, the complier also has to be installed. The compiler file path is set in MPLAB by selecting Project, Set Language Tool Locations. The compiler can then be selected via the Project, Select Language Tool Suite menu option. Browse for the compiler executable file (CCSC.EXE) and select it.

Figure 1.16: Screenshot of MPLAB Project

outbyte.err   The error file provides debugging messages, which are displayed in the Output, Build window after compilation.

outbyte.sym   The symbol map shows the register locations in which the program variables are stored.

outbyte.mcp   This is the MPLAB project information file.

outbyte.mcw   This is the MPLAB workspace information file.

outbyte.pjt   This is the CCS compiler project information file.

1.6   PIC16 Program and Debug

- Programming the chip
- In-circuit debugging
- Design package

Figure 1.17: PICkit2 Demo System Hardware (reproduced by permission of Microchip Inc.)

Once the compiler has produced the hex file, it can be downloaded to the target application board. However, it is generally preferable to test it first by software simulation. This means running the program in a virtual MCU to test its logical function. This can be done within MPLAB (tabular output) or using a third party debugging tool such as Proteus VSM (graphical output). More details on simulation are provided in Appendix C, and VSM interactive simulation is referred to throughout the text to provide circuit schematics and debugging facilities.

Programming

A low-cost programmer available at the time of writing is the Microchip PICkit2 programmer (Figure 1.17). This connects to the USB port of the host PC, with the programming module plugging direct into the target PCB. The six-way in-circuit serial programming (ICSP) connector, between the programmer module and the target board, must be designed into the application circuit. An in-line row of pins provides the programmer connection to the target MCU, as shown in Figure 1.18.

Pin 1 carries the programming voltage (12–14 V) and is connected to pin V_{pp}, which doubles as the MCU reset input, !MCLR. Pin 4 (PGD) carries the program data and pin 5 (PGC), the program clock. Any other circuits connected to these pins must be designed with care, so that they do not interfere with the programmer. The USB output provides the target board power, up to a limit of 500 mA, on pins 2 and 3. If necessary, a separate target board supply must be provided.

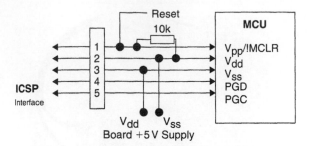

Figure 1.18: ICSP Target Board Connections

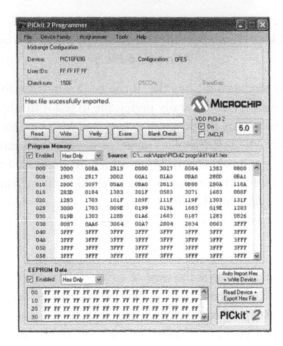

Figure 1.19: PICkit2 Programmer Dialog

Once the hardware is connected up and the programmer drivers loaded, the programming utility window (Figure 1.19) can be opened by running PICkit2.exe file, selected from the Programmer menu. The hex file created by the compiler is imported via the file menu and downloaded using the write button. The target program is run by checking the On box.

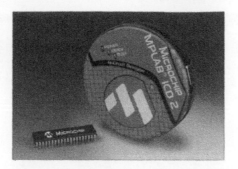

Figure 1.20: Microchip ICD2 Module

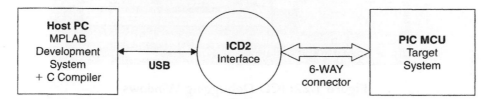

Figure 1.21: ICD2 Program and Debug System

Debugging

If in-circuit debugging is required, the Microchip MPLAB ICD2® in-circuit debugger (Figures 1.20 and 1.21) is recommended. This allows the application program to be tested in real hardware by using the same MPLAB debugging tools used in the simulation mode: source code display, run, stop, step, reset, breakpoints, and variable watch windows. The target system needs its own power supply and an ICD connector.

With power supplied to the target, load the application project files. Select Debugger, Select Tool, MPLAB ICD2. The debug control panel appears with controls to run, step, and reset (Figure 1.22). If the program is recomplied after a change in the source code, the target can be automatically reprogrammed.

Use of breakpoints is generally the most useful debugging technique in C, as it allows complete blocks of assembler to be executed at full speed. These are enabled by right clicking on the source code and indicated by a red marker. Once set, they can be temporarily enabled and disabled. The watch window, selected from the View menu, allows program variable values to be monitored as the program progresses.

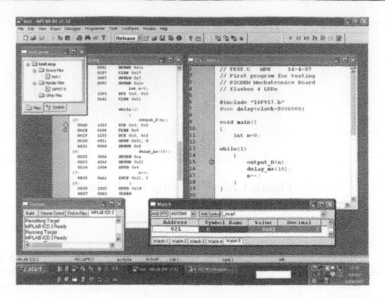

Figure 1.22: ICD Debugging Windows

When debugging has been completed, the chip must be reprogrammed for the final time by selecting Programmer, Select Tool, MPLAB ICD2. Then, hit the Program Target Device button. When done, the program can be stopped and started using the Hold In Reset and Release From Reset buttons. When the ICD pod is disconnected, the program should auto-run in the target system.

Design Package

The components of the ECAD design package used in this book are listed below. The PCB implementation tools are not described further, as they are outside the scope of this programming guide.

- Circuit schematic capture (Proteus ISIS)

- Interactive circuit simulation (Proteus VSM)

- PCB layout design (Proteus ARES)

- PIC development system (Microchip MPLAB)

- PIC C Compiler (Custom Computer Services CCS C)

- PIC programming and in-circuit testing (Microchip ICD2)

Assessment 1

5 points each, total 100

1. List five consumer products that typically include a microcontroller.

2. Identify the five functional elements of a microcontroller.

3. Explain why flash ROM is an important technology in microcontrollers.

4. State five important characteristics of a microcontroller that should be considered when selecting the best part for a given application.

5. Describe briefly the process of program execution in a microcontroller, referring to the role of the program memory, instruction register, program counter, file registers, and working register.

6. State the function of the following registers in the PIC16F877: 02h, 03h, 09h, 89h, 20h.

7. Explain the significance of the following abbreviations in relation to the configuration of the PIC microcontroller: RC, XT, WDT, PUT, NOWRT.

8. Explain the function of the following elements of the PIC I/O circuit: tristate gate, current driver, data direction latch, input data latch, output data latch.

9. A 16-bit PIC hardware timer is driven from the internal clock signal, and the MCU is operating with a 20-MHz crystal. Calculate the preload value required to produce an interrupt every 10 ms.

10. If an analog-to-digital converter has a positive input reference voltage of 2.048 V and is set up as for 8-bit conversion, calculate the resolution of the ADC in millivolts per bit and the output code if the input voltage is 1.000 V.

11. Refer to Figure 1.9, and briefly explain the timer interrupt process and why it is useful.

12. Sketch the RS232 signal that transmits the character X (ASCII code 01011000) on a line operating at ±12V. Indicate the stop and start bits as S and P.

13. Explain the difference between an asynchronous and synchronous data transmission by reference to RS232 and SPI.

14. Explain the difference between hardware and software addressing as used by SPI and I²C.

15. Explain briefly why SPI is generally faster than I²C.

16. A page of plain text contains about 1000 ASCII characters. Estimate the minimum time required to transmit this page over a 9600-baud RS232 link and an SPI line, under the control of an MCU running at 20 MHz, stating any assumptions made.

17. State the function of each of the C project files that have the following extension: C, HEX, COF, LST, and ERR.

18. State the function of the five connections in the PIC in-circuit programming and debugging interface.

19. Study the content of the dissembler window in Figure 1.22, and state the function of the five visible windows.

20. List a minimum set of development system hardware and software components required to create a C application for the PIC microcontroller.

Assignments 1

Assignment 1.1

Download the data book for the PIC16F87X MCUs from www.microchip.com. Study Figure 1.2, the PIC16F877 block diagram. Describe in detail the sequence of events that occurs when the data code for 255_{10} (11111111_2) from a machine code instruction is output to Port D. Refer to the role of the program memory, program counter, instruction register, instruction decoder, file register addressing, internal data bus, and clock. What path must the data follow to get from the program memory to Port C? Describe the setup required in Port C to enable the data byte to be observed on the port pins (Figure 1.4). Refer, if necessary, to *PIC Microcontrollers: An Introduction to Microelectronics* by the author.

Assignment 1.2

Research a list of SPI and I²C peripherals that might be useful in constructing PIC applications. Identify typical memory, interfacing, and sensor chips that use these interfaces and summarize the range of devices available for each interface.

Assignment 1.3

Download and install MPLAB development system from www.microchip.com, and the demo C complier for the PIC16F877 from www.ccsinfo.com. Create the project OUTBYTE as described in Section 1.5. Enter the source code and save in the project folder. Copy the header file into the same folder. Compile the program and view the files created in the folder. Check that the `.hex`, `.lst`, and `.cof` files have been created. Test the program in simulation mode; arrange the MPLAB windows as seen in Figure 1.6 and check that Port C is loaded with the output byte `FFh`. Study the assembler version of the program; note the number of instructions required to implement the C output statement. Reset and step through the program, noting the two phases: initialization and loop. Change the output number in the source code from 255 to 85_{10}, recompile, and run. What is the Port D output now in binary and hex?

Listing 2.1   A Program to Output a Binary Code

```
//  OUTNUM.C Outputs an 8-bit code at Port D in the 16F877 MCU

#include "16F877A.h"      // MCU header file

void main()               // Main block start
{
  output_D(255);          // Switch on outputs
}
```

The essential source code components can be identified. The `include` statement tells the compiler to incorporate the header file for a particular MCU. It provides information about the chip hardware features that the compiler needs to tailor the program. The keywords `void main` indicate the start of the main program block, and the associated braces (curly brackets) enclose the program statements. This program only contains one statement, the function call `output_D(nnn)` that sends a binary code to Port D.

Program Creation

The development process was introduced in Part 1, and further details are provided in Appendices A, B, and C. Briefly, the program project is created as follows:

1. Assuming that MPLAB and CCS C compiler are installed, create a folder for the project files, and an MPLAB project called OUTNUM. Copy the MCU header file 16F877.h from the CCS header file folder to the project folder.

2. Write the program (OUTNUM.C) in the source code edit window of MPLAB, referring to the compiler manual for the correct syntax, and save it in the project folder. Assign the source code and header file in the project window.

3. Build the project (compile and link all files) to create OUTNUM.COF. Correct any syntax and linker errors.

4. Run the program in MPSIM simulation mode. Use the source code debugging window to trace the program execution and the watch window to track the CPU variables. Correct any logical errors.

5. Optionally, the program can be tested in Proteus VSM, which once installed, can be selected from the debugger menu.

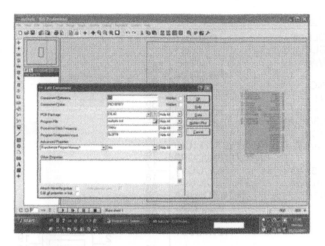

Figure 2.1: ISIS Dialog to Attach the Program

Program Testing

The program could be tested by downloading to a suitable hardware target system, but it is preferable to debug it first in simulation mode, either in MPLAB or, preferably, in Proteus VSM. In the VSM schematic capture and cosimulation module ISIS, the target PIC is selected from the component library and placed on the schematic. The application file OUTNUM.COF previously created by the compiler is attached to it (Figure 2.1) and the schematic saved in the project folder. When the simulation is run, the state of the outputs is indicated by red and blue indicators.

Although not absolutely necessary for program testing in simulation mode, a set of LEDs with their load resistors are attached to Port D, since these are required in the actual hardware to display the outputs (Figure 2.2). No other circuit components or connections are required at this stage, since the simulation runs correctly without a clock circuit. In the real hardware, the clock circuit must be added and !MCLR input tied to V_{dd} (+5V). Here, the clock frequency is set in the MCU properties dialog when the program is attached. To take advantage of the full debugging facilities of MPLAB, Proteus VSM can be run from within MPLAB by installing it in the debug tool menu. For this, a plug-in needs to be downloaded from www.labcenter.co.uk. When selected, the simulator runs in a VSM viewer window (Figure 2.3).

checking the program. This makes subsequent source code debugging and modification easier. The benefits of good layout become more obvious later, when more complex programs are developed.

By tradition, C source code is written mainly in lower case, with upper case used for certain key words.

2.2   PIC16 C Program Basics

- Variables

- Looping

- Decisions

The purpose of an embedded program is to read in data or control inputs, process them, and operate the outputs as required. Input from parallel, serial, and analog ports are held in the file registers for temporary storage and processing; and the results are output later on, as data or a signal. The program for processing the data usually contains repetitive loops and conditional branching, which depends on an input or calculated value.

Variables

Most programs need to process data in some way, and named variables are needed to hold their values. A variable name is a label attached to the memory location where the variable value is stored. When working in assembly language, a register label acts as the variable name and has to be assigned explicitly. In C, the variable label is automatically assigned to the next available location or locations (many variable types need more than 1 byte of memory). The variable name and type must be declared at the start of the program block, so that the compiler can allocate a corresponding set of locations. Variable values are assumed to be in decimal by default; so if a value is given in hexadecimal in the source code, it must be written with the prefix 0x, so that 0xFF represents 255, for example.

A variable called x is used in the program in Listing 2.2, VARI.C. Longer labels are sometimes preferable, such as "output_value," but spaces are not allowed. Only alphanumeric characters (a–z, A–Z, 0–9) and underscore, instead of space, can be used. By default, the CCS compiler is *not* case sensitive, so 'a' is the same as 'A' (even though the ASCII code is different). A limited number of key words in C, such as main and include, must not be used as variable names.

Listing 2.2   Variables

```
/*
  Source code file:         VARI.C
  Author, date, version:    MPB 11-7-07 V1.0
  Program function:         Outputs an 8-bit variable
  Simulation circuit:       OUTBYTE.DSN
  *************************************************************/
#include "16F877A.h"

void main()
{
  int x;                    // Declare variable and type

  x=99;                     // Assign variable value
  output_D(x);              // Display the value in binary
}
```

The variable x is an 8-bit integer with whole number values $0-255_{10}$. The value in binary can be seen when it is output at an 8-bit port. Generally, C integers (int) are stored as 16-bit values, but C for 8-bit microcontrollers uses a default 8-bit integer format. In Program VARI.C, an initial value is assigned to the variable (99), which is then used in the output function. The point here is that the variable value can now be modified without having to change the output function call itself.

In the program, an 8-bit variable x is declared and assigned a value 99 using the "equals" operator. It is then output to Port D using the standard output function.

Looping

Most real-time applications need to execute continuously until the processor is turned off or reset. Therefore, the program generally jumps back at the end to repeat the main control loop. In C this can be implemented as a "while" loop, as in Listing 2.3.

The condition for continuing to repeat the block between the while braces is contained in the parentheses following the while keyword. The block is executed if the value, or result of the expression, in the parentheses is not zero. In this case, it is 1, which means the condition is always true; and the loop repeats endlessly. This program represents in simple form the general structure of embedded applications, where an initialization phase is followed by an endless control loop. Within the loop, the value of x is incremented (x++). The output

---

Listing 2.3   Endless Loop

```
// Source code file:      ENDLESS.C
// Program function:      Outputs variable count

#include "16F877A.h"

void main()
{
  int x;                  // Declare variable

  while(1)                // Loop endlessly
  { output_D(x);          // Display value
    x++;                  // Increment value

  }
}
```

---

therefore appears to count up in binary when executing. When it reaches the maximum for an 8-bit count (11111111 = 255), it rolls over to 0 and starts again.

Decision Making

The simplest way to illustrate basic decision making is to change an output depending on the state of an input. A circuit for this is shown in Figure 2.4, INBIT.DSN. The switch generates an input at RC0 and RD0 provides the test output.

The common keyword for selection in many high level languages is IF. Program IFIN.C (Listing 2.4) has the usual endless "while" loop but contains a statement to switch off Port D initially. The input state is read within the loop using the bit read function input(PIN_C0). This assigns the input value 1 or 0 to the variable x. The value is then tested in the if statement and the output set accordingly. Note that the test uses a double equals to differentiate it from the assignment operator used in the previous statement. The effect of the program is to switch on the output if the input is high. The switch needs to be closed before running to see this effect. The LED cannot be switched off again until the program is restarted.

Loop Control

The program can be simplified by combining the input function with the condition statement as follows:

```
if(input(PIN_C0))output_high(PIN_D0);
```

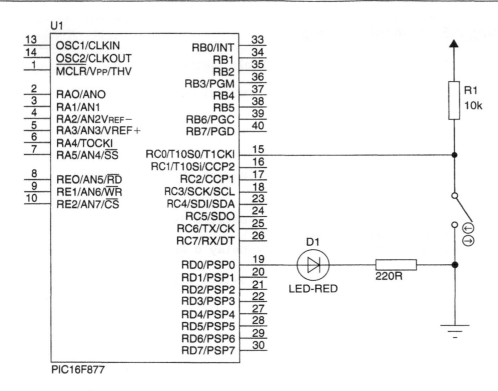

Figure 2.4: INBIT.DSN Test Circuit with Input Switch

Listing 2.4   IF Statement

```
//  IFIN.C  Tests an input
#include "16F877A.h"

void main()
{
  int x;                      // Declare variable
  output_D(0);                // Clear all outputs

  while(1)                    // Loop always
  {
    x = input(PIN_C0);        // Get input state
    if(x==1)output_high(PIN_D0);  // Change output
  }
}
```

Listing 2.7   SIREN Program

```
/*
Source code file:        SIREN.C
Author, date, version:   MPB 11-7-07 V1.0
Program function:        Outputs a siren sound
Simulation circuit:      INBIT.DSN
Compiler:                CCS C Version 4

********************************************************************/
#include "16F877A.h"
#use delay (clock=1000000)

void main()
{
  int step;

  while(1)                         // Keep checking switch
  {
    while(!input(PIN_C0))          // Siren while switch ON
    {
      for(step=0;step<255;step++)  // Loop control
      {
        output_high(PIN_D0);       // Sound sequence
        delay_us(step);
        output_low(PIN_D0);
        delay_us(step);
      }
    }
  }
}
```

The delay is therefore in microseconds. The output is generated when the switch is closed (input C0 low). The delay picks up the incrementing value of "step," giving a longer pulse each time the `for` loop is executed. This causes a burst of 255 pulses of increasing length (reducing frequency), repeating while the input is on. Note that 255 is the maximum value allowed for "step," as it is an 8-bit variable. When run in VSM, the output can be heard via the simulation host PC sound card. Note the inversion of the input test condition using ! = not true.

The header information is now more extensive, as would be the case in a real application. Generally, the more complex a program, the more information is needed in the header. Information about the author and program version and/or date, the compiler version, and the intended target system are all useful. The program description is important, as this summarizes the specification for the program.

Blank Program

A blank program is shown in Listing 2.8, which could be used as a general template. We should try to be consistent in the header comment information, so a standard comment block is suggested. Compiler directives are preceded by hash marks and placed before the main block. Other initialization statements should precede the start of the main control loop. Inclusion of the unconditional loop option `while(1)` assumes that the system will run continuously until reset.

We now have enough vocabulary to write simple C programs for the PIC microcontroller. A basic set of CCS C language components is shown in Table 2.1. Don't forget the semicolon at the end of each statement.

2.3 PIC16 C Data Operations

- Variable types

- Floating point numbers

- Characters

- Assignment operators

A main function of any computer program is to carry out calculations and other forms of data processing. Data structures are made up of different types of numerical and character variables, and a range of arithmetical and logical operations are needed. Microcontroller programs do not generally need to process large volumes of data, but processing speed is often important.

Variable Types

Variables are needed to store the data values used in the program. Variable labels are attached to specific locations when they are declared at the beginning of the program, so the MCU can locate the data required by each operation in the file registers.

Listing 2.8   Program Blank

```
//    Source Code Filename :
//    Author/Date/Version :
//    Program Description :
//    Hardware/simulation :
/////////////////////////////////////////////////////////////

#include "16F877A.h"    // Specify PIC MCU
#use                    // Include library routines

void main()             // Start main block
{
  int                   // Declare global variables

  while(1)              // Start control loop
  {
                        // Program statements
  }
}                       // End main block
```

Table 2.1: A Basic Set of CCS C Source Code Components

C Compiler Directives	
`#include source files`	Include source code or header file
`#use functions(parameters)`	Include library functions
C Program Block	
`main(condition){statements}`	Main program block
`while(condition){statements}`	Conditional loop
`if(condition){statements}`	Conditional sequence
`for(condition){statements}`	Preset loop
CCS C Library Functions	
`delay_ms(nnn)`	Delay in milliseconds
`delay_us(nnn)`	Delay in microseconds
`output_x(n)`	Output 8-bit code at Port X
`output_high(PIN_nn)`	Set output bit high
`output_low(PIN_nn)`	Set output bit low
`input(PIN_nn)`	Get input

Table 2.2: Range of Integer Variables

Name	Type	Minimum	Maximum	Range
int1	1 bit	0	1	$1 = 2^0$
unsigned int8	8 bits	0	255	$256 = 2^8$
signed int8	8 bits	-127	$+127$	$256 = 2^8$
unsigned int16	16 bits	0	65535	$65536 = 2^{16}$
signed int16	16 bits	-32767	$+32767$	$65536 = 2^{16}$
unsigned int32	32 bits	0	4294967295	$4294967296 = 2^{32}$
signed int32	32 bits	-2147483647	$+2147483647$	$4294967296 = 2^{32}$

Integers

We have seen the integer (whole number) variable in use. In the 8-bit MCU, the default type is an unsigned 8-bit number, giving a range of values of 0–255. This obviously is inadequate for many purposes, so 16- and 32-bit integer types are also needed (see Table 2.2). The range of a number is determined by the number of different binary codes that can be represented. If n is the number of bits, 2^n different codes are possible. As 0 must be included, the highest number is $2^{(n-1)}$. Hence, the 16-bit unsigned integer has the range 0–65535 ($2^{16} - 1$) and the 32 bit 0–4294967295 ($2^{32} - 1$). There is also a 1-bit type for bit storage.

Signed Integers

The signed integer uses the most significant bit (MSB) as the sign bit, so the range is accordingly reduced by half. MSB = 0 represents a positive number, MSB = 1 indicates a negative number. Therefore, the range for a 16-bit signed integer is -32767 to $+32767$. The sign bit must be processed separately to get the right answer from a calculation.

Floating Point

Integers can represent only a limited range of numbers, with a precision of ±0.5. Therefore, the floating point (FP) type should be used for many calculations, particularly those with a fractional result. The 32-bit FP format can represent decimal numbers from about 10^{-39} to 10^{+38}, with a precision of about 10^{-7} (±0.0000001). The number is stored in exponential format, as used in a standard calculator. Twenty-three bits are used for the significant digits, called the *mantissa*. Eight bits are used for the exponent part and one for the sign. The IEEE standard form has the sign bit as the MSB, but Microchip and CCS use a slightly more logical form, where the sign bit is the MSB of the third byte, leaving the exponent to be represented by the complete high byte (Table 2.3).

The significant figures of the floating point number (mantissa) are represented by a positive fractional binary number whose value is between 0 and 1. As in any binary number, the weighting of the 23 bits is a power of 2 series but fractional, that is, ½, ¼, ⅛, ¹⁄₁₆, ¹⁄₃₂, ¹⁄₆₄, …, ½²³. The final fraction represents the resolution of the format, that is, the smallest step in the number sequence:

$$1/2^{23} = 1/8388608 \approx 0.0000001 = 10^{-7}$$

Hence, 32-bit floating point numbers are precise to about seven decimal places. The final result can therefore be quoted to six decimal places, assuming that rounding errors are not significant.

An example of a floating point number is given in Table 2.4. Its value can be determined by following the process of conversion that comes next, which is the easiest way to describe the FP format.

The 32-bit FP number given is

```
1000 0011 1101 0010 0000 0000 0000 0000
```

Table 2.3: Microchip/CCS Floating Point Number Format

Exponent	Sign	Mantissa
eeee eeee	s	mmm mmmm mmmm mmmm mmmm mmmm
8 bits	1 bit	23 bits

Table 2.4: Example of 32-Bit Floating Point Number Format

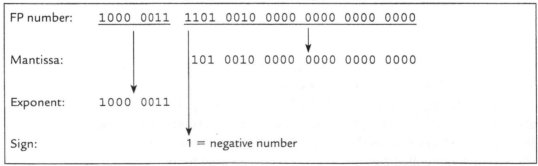

FP number: 1000 0011 1101 0010 0000 0000 0000 0000

Mantissa: 101 0010 0000 0000 0000 0000

Exponent: 1000 0011

Sign: 1 = negative number

The mantissa is the low 23 bits, and the set bit weighting gives the value

$$1/2 + 1/8 + 1/64 = 0.5 + 0.125 + 0.015625 = 0.640625$$

Then, 1 is added to shift the decimal part into the range between 1.9999999 and 1.000000:

$$\text{Decimal number} = 1.640625$$

$$\text{Signed result} = -1.640625$$

The exponent is given by the high byte: $1000\ 0011 = 131_{10}$

This includes an offset of 127 to allow for positive and negative exponents, so we subtract 127 to obtain the corrected exponent: $131 - 127 = +4$

The multiplier value is then calculated from the binary exponent: $2^{+4} = 16$

The final value is found by multiplying this by the mantissa result:

$$16 \times -1.640625 = -26.25$$

The range of numbers that can be represented by the FP format can be estimated from the exponent range:

$$\text{Minimum exponent value:} \quad 2^{-127} \approx 10^{-39}$$

$$\text{Maximum exponent value:} \quad 2^{128} \approx 10^{+38}$$

This is adequate for most purposes. The disadvantage of this format is there are always slight rounding errors; so if an integer is converted to a FP number and back, it no longer is exact. This is illustrated in Figure 2.5, where integer variables have been assigned their maximum values in a demo program and are displayed in the watch window after running in MPSIM. The integers are correct, but the discrepancy due to rounding errors between the working value of the floating point number and the original can be seen to be 12.3456793 − 12.3456789 = 0.0000004.

One advantage of C is that the exact method of calculation is normally concealed within the built-in functions and operations. However, we still need to use the most appropriate numerical format, because the C compiler does not tell us if the right answer is obtained from any given calculation. This is where simulation is useful in real-time applications—we can check that the answers are correct before they are used to modify control outputs in real hardware. The integer types and ranges available in CCS C are shown in Table 2.1.

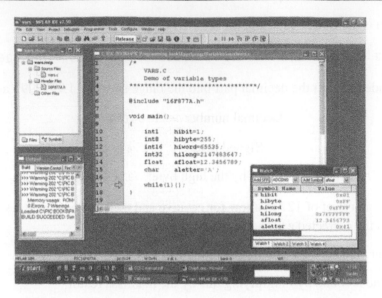

Figure 2.5: Variable Types Demo Program Screenshot

Character Variable

Text characters are generally represented by ASCII codes (Table 2.5). The basic set of 7-bit characters includes the upper and lower case letters and the numerals and punctuation marks found on the standard computer keyboard. For example, capital (upper case) A is 1000001 (65_{10}). The numeric characters run from 0x30 (0) to 0x39 (9), so to convert to the actual number from ASCII, simply subtract 0x30. The character variable is indicated in C source code in single quotes. For example the statement answer = 'Y'; will assign the value 0x59 to the variable 'answer'.

Assignment Operations

A range of arithmetic and logic operations are needed where single or paired operands are processed. The result is assigned to one of the operand variables or a third variable.

Integers can be used for simple unsigned arithmetic operations, giving an exact result. However, in general, floating point numbers must be used for signed calculations, but remember there will be small errors. Logical operations must use integers, as the numbers are processed bit by bit. A complete set of operators is listed in Table 2.6.

Table 2.5: The 7-Bit ASCII Codes

Low Bits	High Bits					
	010	011	100	101	110	111
0000	Space	0	@	P	`	p
0001	!	1	A	Q	a	q
0010	"	2	B	R	b	r
0011	#	3	C	S	c	s
0100	$	4	D	T	d	t
0101	%	5	E	U	e	u
0110	&	6	F	V	f	v
0111	'	7	G	W	g	w
1000	(	8	H	X	h	x
1001	)	9	I	Y	i	y
1010	*	:	J	Z	j	z
1011	+	;	K	[	k	{
1100	,	<	L	\	l		
1101	-	=	M	]	m	}
1110	.	>	N	^	n	~
1111	/	?	O	_	o	Del

Figure 2.6 shows the output of a test program that carries out some sample operations. The results are shown in a watch window after running the program in MPSIM. The 8-bit integer operations give the correct output while the result is in range. The product of the multiplication (mulbyte) is clearly incorrect, while the result of the integer division (divbyte) is truncated. Floating point calculations are required in this case. The floating point results show nine significant figures, but only four are valid for the addition and subtraction, seven for the multiplication, and the division result is also correct only to seven figures.

Table 2.6: Arithmetic and Logical Operations

Operation	Operator	Description	Source Code	Example	Result
Single operand					
Increment	++	Add 1 to integer	`result = num1++;`	0000 0000	0000 0001
Decrement	--	Subtract 1 from integer	`result = num1--;`	1111 1111	1111 1110
Complement	~	Invert all bits of integer	`result = ~num1;`	0101 0010	1010 1101
Arithmetic operation					
Add	+	Integer or float	`result = num1+num2;`	0000 1010 +0000 0111	0001 0001
Subtract	−	Integer or float	`result = num1-num2;`	0000 1010 −0000 0011	0000 0111
Multiply	*	Integer or float	`result = num1*num2;`	0000 1010 *0000 0011	0001 1110
Divide	/	Integer or float	`result = num1/num2;`	0000 1100 /0000 0011	0000 0100
Logical operation					
Logical AND	&	Integer bitwise	`result = num1&num2;`	1001 0011 &0111 0001	0001 0001
Logical OR	|	Integer bitwise	`result = num1|num2;`	1001 0011 |0111 0001	1111 0011
Exclusive OR	^	Integer bitwise	`result = num1^num2;`	1001 0011 ^0111 0001	1110 0010

Conditional Operations

Where a logical condition is tested in a while, if, or for statement, relational operators are used. One variable is compared with a set value or another variable, and the block is executed if the condition is true. The conditional operators are shown in Table 2.7. Note that double equals is used in the relational test to distinguish it from the assignment operator.

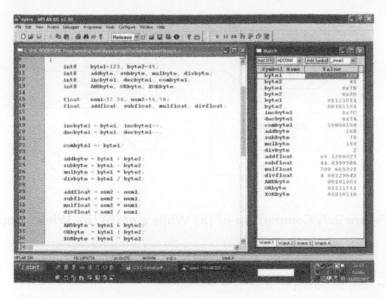

Figure 2.6: Results of Sample Arithmetic and Logic Operations in MPLAB Program Simulation

Table 2.7: Conditional Operators

Operation	Symbol	Example
Equal to	==	if(a==0) b=b+5;
Not equal to	!=	if(a!=1) b=b+4;
Greater than	>	if(a>2) b=b+3;
Less than	<	if(a<3) b=b+2;
Greater than or equal to	>=	if(a>=4) b=b+1;
Less than or equal to	<=	if(a<=5) b=b+0;

Sometimes, a conditional test needs to combine tests on several values. The tests can be compounded by using logical operators, as follows:

AND condition: if((a>b)&&(c=d))...

OR condition: if((a>b)||(c=d))...

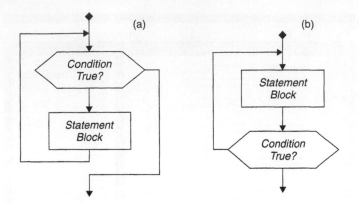

Figure 2.7: Comparison of (a) While and (b) Do..While Loop

2.4   PIC16 C Sequence Control

- While loops

- Break, continue, goto

- If, else, switch

Conditional branching operations are a basic feature of any program. These must be properly organized so that the program structure is maintained and confusion avoided. The program then is easy to understand and more readily modified and upgraded.

While Loops

The basic `while(condition)` provides a logical test at the start of a loop, and the statement block is executed only if the condition is true. It may, however, be desirable that the loop block be executed at least once, particularly if the test condition is affected within the loop. This option is provided by the `do..while(condition)` syntax. The difference between these alternatives is illustrated in Figure 2.7. The WHILE test occurs before the block and the DO WHILE after.

The program DOWHILE shown in Listing 2.9 includes the same block of statements contained within both types of loop. The WHILE block is not executed because the loop control variable has been set to 0 and is never modified. By contrast, 'count' is incremented within the DO WHILE loop before being tested, and the loop therefore is executed.

Listing 2.9   DOWHILE.C Contains Both Types of 'While' Loop

```c
//  DOWHILE.C
//  Comparison of WHILE and DO WHILE loops

#include "16F877A.H"

main()
{
  int outbyte1=0;
  int outbyte2=0;
  int count;
// This loop is not executed .............
  count=0;
  while (count!=0)
  {    output_C(outbyte1);
       outbyte1++;
       count--;
  }
// This loop is executed..................
  count=0;
  do
  {    output_C(outbyte2);
       outbyte2++;
       count--;
  } while (count!=0);

  while(1){};
}
```

Break, Continue, and Goto

It may sometimes be necessary to break the execution of a loop or block in the middle of its sequence (Figure 2.8). The block must be exited in an orderly way, and it is useful to have the option of restarting the block (continue) or proceeding to the next one (break). Occasionally, an unconditional jump may be needed, but this should be regarded as a last resort, as it tends to threaten the program stability. It is achieved by assigning a label to the jump destination and executing a `goto..label`.

The use of these control statements is illustrated in Listing 2.10. The events that trigger break and continue are asynchronous (independent of the program timing) inputs from external switches, which allows the counting loop to be quit or restarted at any time.

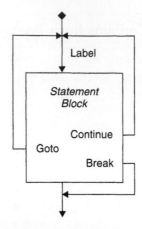

Figure 2.8: Break, Continue, and Goto

Listing 2.10   Continue, Break, and Goto

```
//  CONTINUE.C
//  Continue, break, and goto jumps

#include "16F877A.H"
#use delay(clock=4000000)

main()
{
  int outbyte;

  again: outbyte=0;                  // Destination of goto

  while(1)
  {
    output_C(outbyte);               // Foreground operation
    delay_ms(10);
    outbyte++;                       // Increments Port C

    if (!input(PIN_D0)) continue;    // Skip other tests if input 0
                                     // low
    if (!input(PIN_D1)) break;       // Terminate loop if input 1 low
    delay_ms(100);                   // Debounce inputs
    if (outbyte==100) goto again;    // Restart at 100
  }
}
```

The goto again is triggered by the count reaching a set value, which could be better achieved by using the While condition. In a more complex program, exiting a function in this way risks disrupting program control, since the function is not properly terminated. The significance of this should become clearer when functions are analyzed later.

If..Else and Switch..Case

We have seen the basic if control option, which allows a block to be executed or skipped conditionally. The else option allows an alternate sequence to be executed, when the if block is skipped. We also need a multichoice selection, which is provided by the switch..case syntax. This tests a variable value and provides a set of alternative sequences, one of which is selected depending on the test result.

These options are illustrated in flowchart form in Figures 2.9 and 2.10, and the if.. else and switch..case syntax is shown in Listing 2.11. The control statement switch(variable) tests the value of the variable used to select the option block. The keyword case n: is used to specify the value for each option. Note that each option block must be terminated with break, which causes the remaining blocks to be skipped. A default block is executed if none of the options is taken.

The same effect can be achieved using if..else, but switch..case provides a more elegant solution for implementing multichoice operations, such as menus. If the case options comprise more than one statement, they are best implemented using a function block call, as explained in the next section.

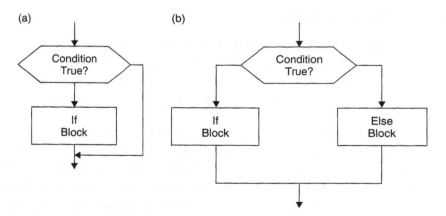

Figure 2.9: Comparison of (a) If and (b) If..Else

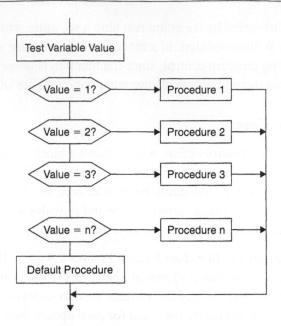

Figure 2.10: Switch..Case Branching Structure

2.5   PIC16 C Functions and Structure

- Program structure

- Functions, arguments

- Global and local variables

The structure of a C program is created using functions (Figure 2.11). This is a block of code written and executed as a self-contained process, receiving the required parameters (data to be processed) from the calling function and returning results to it. Main() is the primary function in all C programs, within which the rest of the program is constructed.

When running on a PC, main() is called by the operating system, and control is returned to the OS when the C program is terminated. In the microcontroller, main() is simply used to indicate the start of the main control sequence, and more care needs to be taken in terminating the program. Normally, the program runs in a continuous loop, but if not, the final statement should be while(1);, which causes the program to wait and prevents the program running into undefined locations following the application code.

Listing 2.11   Comparison of Switch and If..Else Control

```
//   SWITCH.C
//   Switch and if..else sequence control
//   Same result from both sequences
//////////////////////////////////////////////////////////////
#include "16F877A.h"

void main()
{
  int8 inbits;

  while(1)
  {
    inbits = input_D();            // Read input byte

    switch(inbits)                 // Test input byte
    {
      case 1: output_C(1);         // Input = 0x01, output = 0x01
              break;               // Quit block
      case 2: output_C(3);         // Input = 0x02, output = 0x03
              break;               // Quit block
      case 3: output_C(7);         // Input = 0x03, output = 0x07
              break;               // Quit block
      default:output_C(0);         // If none, output=0x00
    }

    if (input(PIN_D0)) output_C(1); // This block has same effect
    if (input(PIN_D1)) output_C(2);
    if (input(PIN_D0) && input(PIN_D1)) output_C(7);
    else output_C(0);
  }
}
```

We have already seen built-in functions such as `input(PIN_D0)` and `output_C(255)`, which read and write the ports. Function "arguments," given in the parentheses, allow function parameters to be passed to the function block, in this case specifying the port or pin to be accessed. Another example is `delay_ms(100)`, which passes the required delay time to the delay function.

In this case, the function code must be called up explicitly with the `#use delay (clock=4000000)` directive. This tells the compiler to include the delay library functions, allowing the system clock to be specified at the same time, so that the correct delays can be calculated.

Listing 2.13   Passing a Parameter to a Function

```
//  FUNC2.C
//  Uses global variables only
////////////////////////////////////////////////////////////
#include "16F877A.H"

int8 outbyte=1;    // Declare global variables
int16 n,count;

  void out()     ////////////////////// Function to run output count
  {
    while(outbyte!=0)
     {  output_C(outbyte);
        outbyte++;
        for(n=1;n<count;n++);    // Use global value for count
     }
  }
main()       ////////////////////// Main block
{
  count=2000;    // Set variable value
  out();         // Call function
  while(1);      // Wait for reset
}
```

variables are used. The function `out()` runs a binary count, which is stopped when a switch on pin D0 is closed. This value is then returned to the main program and displayed. Variable n is local to function `out()` and is declared within the function. Variable t is also local but receives its value from the variable count in the calling routine. The value is transferred between the argument in the function call (`count`) and the argument of the function declaration (`int16 t`). Note that the local integer type must be declared in the function declaration. The function also returns a value `outbyte` to the main block. This is displayed at Port C in the main routine.

2.6   PIC16 C Input and Output

- RS232 serial data

- Serial LCD

- Calculator and keypad

Listing 2.14   Using Local Variables in Functions

```
// FUNC3.C
// Uses local variables
//////////////////////////////////////////////////////////////

#include "16F877A.H"

int8 outbyte=1;                // Declare global variables
int16 count;

    int out(int16 t)    ////////////// Declare argument types
    {
      int16 n;                 // Declare local variable

      while (input(PIN_D0))  // Run at speed t
      {  outbyte++;
         for(n=1;n<t;n++);
      }
      return outbyte;          // Return output when loop stops
    }
main()    /////////////////////////////////////////////////
{
  count=50000;
  out(count);                  // Pass count value to function
  output_C(outbyte);           // Display returned value
  while(1);
}
```

If an electronic gadget has a small alphanumeric LCD, the chances are that it is a microcontroller application. Smart card terminals, mobile phones, audio systems, coffee machines, and many other small systems use this display. The LCD we use here has a standard serial interface, and only one signal connection is needed. The signal format is RS232, a simple low-speed protocol that allows 1 byte or character code to be sent at a time. The data sequence also includes start and stop bits, and simple error checking can be applied if required. The PIC 16F877, in common with many microcontrollers, has a hardware RS232 port built in. Further details of RS232 are found elsewhere in this book.

Serial LCD

CCS C provides an RS232 driver routine that works with any I/O pin (that is, the hardware port need not be used). This is possible because the process for generating the RS232 data frame is not too complex and can be completed fast enough to generate the signal in real time. At the standard rate of 9600 baud, each bit is about 100 µs long, giving an overall frame time of about 1 ms. The data can be an 8-bit integer or, more often, a 7-bit ASCII character code. This method of transferring character codes via a serial line was originally used in mainframe computer terminals to send keystrokes to the computer and return the output—that is how long it's been around.

In this example, the LCD receives character codes for a 2-row × 16-character display. The program uses library routines to generate the RS232 output, which are called up by the directive #use RS232. The baud rate must be specified and the send (TX) and receive (RX) pins specified as arguments of this directive. The directive must be preceded by a #use delay, which specifies the clock rate in the target system. The LCD has its own controller, which is compatible with the Hitachi 44780 MCU, the standard for this interface.

When the system is started, the LCD takes some time to initialize itself; its own MCU needs time to get ready to receive data. A delay of about 500 ms should be allowed in the main controller before attempting to access the LCD. A basic program for driving the LCD is shown in Listing 2.15.

Characters are sent using the function call putc(code), whose argument is the ASCII code for the character; the ASCII table given previously (Table 2.5) lists the available codes. Note that the codes for '0' to '9' are 0x30 to 0x39, so conversion between the code and the corresponding number is simple. Characters for display can be defined as 'A' to 'Z' and so on, in single quotes, in the program.

The character is then replaced by its code by the compiler. The display also needs control codes, for example, to clear the display and reset the cursor to the start position after characters have been printed. These are quoted as an integer decimal and sent as binary. Each control code must be preceded by the code 254 (1111 1110) to distinguish it from data. The code to start the second line of the display is 192. The display reverts automatically to data mode after any control code. A basic set of control codes is identified in Table 2.8.

In the example program LCD.C, the sample character 'acap' is upper case 'A', ASCII code = 1000001 = 65_{10}. If a string of fixed characters are to be displayed, the form printf("sample text") can be used. The meaning of the function name is "print formatted." We often need to insert a variable value within fixed text; in this case, a format code is placed within the display text, and the compiler replaces it with the value of the variable, which is quoted at the end of the printf statement. The code %d means display the variable value as an integer decimal number, %c means display the ASCII character corresponding to the number. Multiple values can be inserted in order, as seen in program LCD.C. A summary of formatting codes is shown in Table 2.9.

Listing 2.15   Serial LCD Operation

```
//  LCD.C
//  Serial LCD test-send character using putc() and printf()
//////////////////////////////////////////////////////////////

#include "16F877A.h"
#use delay(clock=4000000)
#use rs232(baud=9600, xmit=PIN_D0, rcv=PIN_D1)     // Define speed
                                                   //  and pins

void main()
{
  char acap='A';         // Test data

  delay_ms(1000);        // Wait for LCD to wake up
  putc(254); putc(1);    // Home cursor
  delay_ms(10);          // Wait for LCD to finish

  while(1)
  {
    putc(acap);                                  // Send test character
    putc(254); putc(192); delay_ms(10);          // Move to second row
    printf("ASCII %c CHAR %d ",acap,acap);       // Send test data again
    while(1);
  }
}
```

Table 2.8: Essential Control Codes for Serial 2x16 LCD

Code	Effect
254	Switch to control mode
followed by	
00	Home to start of row 1
01	Clear screen
192	Go to start of row 2

Listing 2.16 shows the program FLOAT.C, which illustrates how different variable types are displayed, as well as showing the range of each type. Each variable type is output in turn to the display. The general form of the format code is %nt, where n is the number of significant figures to be displayed and t is the output variable type. The number of

Table 2.9: Output Format Codes

Code	Displays
%d	Signed integer
%u	Unsigned integer
%Lu	Long unsigned integer (16 or 32 bits)
%Ls	Long signed integer (16 or 32 bits)
%g	Rounded decimal float (use decimal formatting)
%f	Truncated decimal float (use decimal formatting)
%e	Exponential form of float
%w	Unsigned integer with decimal point inserted (use decimal formatting)
%X	Hexadecimal
%LX	Long hex
%c	ASCII character corresponding to numerical value
%s	Character or string

decimal places printed can also be specified for floating point numbers; for example, %5.3d displays a decimal number with five significant digits and three decimal places.

Keypad and Calculator

A simple calculator application demonstrates the use of the LCD and a keypad, as well as some numerical processing.

A matrix keypad provides a simple data entry device for microcontroller systems. The keys are connected in rows and columns, such that pressing a button connects a row to a column. The required connections are shown in Figure 2.12. The rows, labeled A, B, C, and D, are connected as outputs at Port B, avoiding the programming pins. The columns, labeled 1, 2, 3, and 4, are connected as inputs on Port D and are pulled up to +5V by 10-k resistors. A serial LCD, described previously, is driven from pin 7 of Port D.

To read the keypad, each row is set low in turn and the state of the inputs tested. If no button is pressed, all the inputs remain high. When a key is operated, a low on that

Listing 2.16 Formatted Variable Output to a Serial Display

```c
/*  FLOAT.C MPB 4-3-07
    Displays variable types and ranges
****************************************************/

#include "16F877A.h"

#use delay(clock=4000000)
#use rs232(baud=9600, xmit=PIN_D0, rcv=PIN_D1)

int1 minbit=0, maxbit=1;
signed int8 minbyte=-127, maxbyte=127;
signed int16 minword=-32767, maxword=32767;
signed int32 minlong=-2147483647, maxlong=2147483647;
float testnum=12345.6789;

void main()
{
  delay_ms(1000);        // Wait for LCD to wake
  putc(254); putc(1);    // Home cursor
  delay_ms(10);          // Wait for LCD to do

  while(1)
  {
    printf("Bit:%d or %d",minbit, maxbit); delay_ms(1000);
    putc(254); putc(1); delay_ms(10);

    printf("Byte %d to %d",minbyte, maxbyte);    delay_ms(1000);
    putc(254); putc(1);  delay_ms(10);

    printf("Word %Ld",minword); putc(254); putc(192);
    delay_ms(10); printf(" to %Ld",maxword); delay_ms(1000);
    putc(254); putc(1); delay_ms(10);

    printf("Long %Ld",minlong); putc(254); putc(192);
    delay_ms(10); printf(" to %Ld",maxlong); delay_ms(1000);
    putc(254); putc(1); delay_ms(10);

    printf("Float %5.4g",testnum); putc(254); putc(192);
    delay_ms(10); printf("or %e", testnum); delay_ms(1000);
    putc(254); putc(1); delay_ms(10);
  }
}
```

row is detected on the column input for that key, which allows a corresponding code to be generated. This is a binary number or ASCII code, as required by the particular application. Program CALC.C (Listing 2.17) runs on this hardware and implements a simple calculator with limited range.

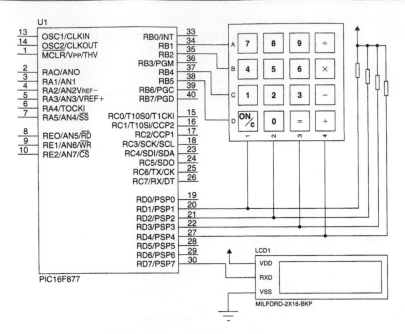

Figure 2.12: Calculator Schematic

Listing 2.17   Calculator Program

```
/*
    Source Code Filename:     CALC.C
    Author/Date/Version:      MPB 21-12-07
    Program Description:      Calculator demo program
    Hardware/simulation :     CALC.DSN

***********************************************************************/
#include "16F877A.h"
#use delay(clock=4000000)
#use rs232(baud=9600,xmit=PIN_D7,rcv=PIN_D0)

// Declare variables *******************************************

int akey, keynum, opcode, numofdigs, start;
int32 num1, num2, result, rem1, rem2, rem3, rem4;
int32 hunsdig, tensdig, onesdig;
int32 hunthous, tenthous, thous, hunds, tens, ones;
```

```
// Declare functions *****************************************************
void scankey();  // Read keypad
void makenum();  // Construct input decimal from keys

// MAIN PROGRAM: Get numbers & calculate *********************************
void main()
{
  for(;;)
  {
    // Get numbers ........................................................

    delay_ms(500); putc(254); putc(1); delay_ms(10);  // Clear display
    numofdigs=onesdig=tensdig=hunsdig=0; akey=0x30;

    do
    {  scankey();                                     // Get first number
       putc(akey);
       if((akey>=0x30)&&(akey<=0x39)) makenum();
    } while((akey>=0x30)&&(akey<=0x39));

    num1=(onesdig+(tensdig*10)+(hunsdig*100));        // Calculate it
    opcode=akey;

    numofdigs=onesdig=tensdig=hunsdig=0; akey=0x30;  // Get second number
    do
    {  scankey();
       putc(akey);
       if((akey>=0x30)&&(akey<=0x39)) makenum();
    } while((akey<=0x30)&&(akey<=0x39));

    num2=(onesdig+(tensdig*10)+(hunsdig*100));        // Calculate it

    // Calculate result...................................................
    if(opcode==0x2F) result=num1/num2;
    if(opcode==0x2A) result=num1*num2;
    if(opcode==0x2D) result=num1-num2;
    if(opcode==0x2B) result=num1+num2;

    //Calc result digits.................................................
    hunthous=result/100000; rem1=result-(hunthous*100000);
    tenthous=rem1/10000; rem2=rem1-(tenthous*10000);
    thous=rem2/1000; rem3=rem2-(thous*1000);
    hunds=rem3/100; rem4=rem3-(hunds*100);
    tens=rem4/10; ones=rem4-(tens*10);

    // Display digits....................................................

    start=0;
    if(hunthous!=0){putc(hunthous+0x30);start=1;}
```

The data in a C program may be most conveniently handled as sets of associated variables. These occur more frequently as the program data becomes more complex, but only the basics are mentioned here.

Arrays

Arrays are sets of variable values having the same type and meaning. For example, each word in a text file is stored as a character array, a sequence of ASCII codes. This is also referred to as a *string*. A numerical array might be a sequence of voltage readings from an analog input in a test system or controller. The program ARRAYS.C (Listing 2.18) shows how they can be created and displayed. The arrays are declared using a collective name

---

Listing 2.18   Numerical and Character Arrays

```
// ARRAYS.C
// Demo of numerical and string arrays
// Attach ARRAYS.COF to LCD.DSN to display
//////////////////////////////////////////////////////////

#include "16F877A.h"
#use delay(clock=4000000)
#use rs232(baud=9600, xmit=PIN_D0, rcv=PIN_D1)

main()
{
  int8 aval=0, n;     // Declare single variables
  int8 anum[10];      // Declare integer array
  char astring[16];   // Declare character array

// Start LCD............................................
  delay_ms(1000);
  putc(254); putc(1); delay_ms(10);

// Assign data to arrays................................
  for ( n=0; n<10; n++ ) { anum[n]=aval; aval++; }
  strcpy(astring,"Hello!");

// Display data.........................................
  for ( n=0; n<10; n++ ) printf("%d",anum[n]);
  putc(254); putc(192); delay_ms(10);
  puts(astring);

  while(1);    // Wait
}
```

and subscript placeholder (anum[10] and astring[16]), which instructs the compiler to allocate a suitable set of locations in RAM. The variable type declaration determines how many locations per value are needed.

The numerical array values are initialized using a for loop; a variable n, which increments from 0 to 9, is used as loop counter and also as the array index value. The character array values are assigned using the function strcpy() (string copy). Its arguments are the target array name astring and the text in double quotes, which is copied to the array. The end of the string is automatically terminated by a zero value, creating a "null terminated string." This allows the end of the message to be easily detected by a receiving device.

The numerical data are displayed on our 16x2 LCD using printf(), again using a for loop. The string is output in a different manner; the puts() (put string) function is simpler than printf() and avoids the need to output each character separately, using putc(). However, printf() is still more convenient for displaying a fixed string.

Table 2.10 shows the contents of the RAM file registers after the program ARRAYS has executed. It can be seen that the numerical array data has been allocated to locations 0x21 to 0x2A inclusive in the GPRs, with the character data in locations 0x2D to 0x32 inclusive. The characters are displayed in the right column, converted from ASCII. The single integers are seen in the locations 0x2B and 0x2C (final value 0x0A). The data bytes can be accessed directly in these locations using indirect addressing operators.

Indirect Addressing Operators

C provides various ways of manipulating data in memory. Since there always seems to be several ways to get the same result, this can be confusing for the beginner. If a variable

Table 2.10: MPLAB Display of Array Data in File Register

Address	00	01	02	03	04	05	06	07	08	09	0A	0B	0C	0D	0E	0F	ASCII
000	--	00	38	1C	00	00	00	00	01	00	00	00	00	00	00	00	--.8..... ....
010	00	00	00	00	00	00	00	00	00	00	00	00	00	00	00	00	....
020	00	00	01	02	03	04	05	06	07	08	09	0A	0A	48	65	6C	....Hel
030	6C	6F	21	00	00	00	00	00	00	00	00	00	00	32	00	18	lo!..... .....2..
040	20	20	20	39	14	0A	00	00	00	00	00	00	00	00	00	00	9....... ....

Compiler directives are typically used at the top of the program to set up compiler options, control project components, define constant labels, and so on before the main program is created. They are preceded by the hash symbol to distinguish them from other types of statements and do not have a semicolon to end the line.

Program Directives

Examples using the directives encountered thus far follow—refer to the compiler reference manual for the full range of options.

```
#include "16F877A.h"
```

The include directive allows source code files to be included as though they had been typed in by the user. In fact, any block of source code can be included in this way, and the directive can thus be used to incorporate previously written reusable functions. The header file referred to in this case provides the information needed by the complier to create a program for a specific PIC chip.

```
#use delay(clock=4000000)
```

The 'use' directive allows library files to be included. As can be seen, additional operating parameters may be needed so that the library function works correctly. The clock frequency given here needs to be specified so that both software and hardware timing loops can be correctly calculated.

```
#use rs232(baud=9600, xmit=PIN_D0, rcv=PIN_D1)
```

In this directive, the parameters set the RS232 data (baud) rate and the MCU pins to be used to transmit and receive the signal. This software serial driver allows any available pin to be used.

Header File

A selection of the more commonly used directives are seen in the processor header file, which must be included in every program. The file 16F877A.H is reproduced in full in Listing 2.19.

The device directive selects the target processor, and can be followed by various options. One that we use later is ADC=8, which sets the resolution of the analog input conversion.

Listing 2.19   Header File 16F877A.H

```
//////// Standard Header file for the PIC16F877A device /////////
#device PIC16F877A
#nolist
//////// Program memory: 8192x14 Data RAM: 367 Stack: 8
//////// I/O: 33 Analog Pins: 8
//////// Data EEPROM: 256
//////// C Scratch area: 77 ID Location: 2000
//////// Fuses: LP,XT,HS,RC,NOWDT,WDT,NOPUT,PUT,PROTECT,DEBUG,NODEBUG
//////// Fuses: NOPROTECT,NOBROWNOUT,BROWNOUT,LVP,NOLVP,CPD,NOCPD,WRT_50%
//////// Fuses: NOWRT,WRT_25%,WRT_5%
////////
//////////////////////////////////////////////////////////////////////
//
// Discrete I/O Functions: SET_TRIS_x(), OUTPUT_x(), INPUT_x(),
//                         PORT_B_PULLUPS(), INPUT(),
//                         OUTPUT_LOW(), OUTPUT_HIGH(),
//                         OUTPUT_FLOAT(), OUTPUT_BIT()
//
// Constants used to identify pins in the above are:
#define PIN_A0 40     // Register 05, pin 0 (5x8)+0=40
#define PIN_A1 41     // Register 05, pin 1 (5x8)+1=41
#define PIN_A2 42     // Register 05, pin 2 (5x8)+2=42
#define PIN_A3 43     // Register 05, pin 3 etc
#define PIN_A4 44     // Register 05, pin 4
#define PIN_A5 45     // Register 05, pin 5

#define PIN_B0 48     // Register 06, pin 0 (6*8)+0=48
#define PIN_B1 49     // Register 06, pin 1 etc
#define PIN_B2 50     // Register 06, pin 2
#define PIN_B3 51     // Register 06, pin 3
#define PIN_B4 52     // Register 06, pin 4
#define PIN_B5 53     // Register 06, pin 5
#define PIN_B6 54     // Register 06, pin 6
#define PIN_B7 55     // Register 06, pin 7

#define PIN_C0 56     // Register 07, pin 0 (7*8)+0=56
#define PIN_C1 57     // Register 07, pin 1 etc
#define PIN_C2 58     // Register 07, pin 2
#define PIN_C3 59     // Register 07, pin 3
#define PIN_C4 60     // Register 07, pin 4
#define PIN_C5 61     // Register 07, pin 5
#define PIN_C6 62     // Register 07, pin 6
#define PIN_C7 63     // Register 07, pin 7
```

```
#define PIN_D0 64          // Register 08, pin 0 (8*8)+0=64
#define PIN_D1 65          // Register 08, pin 1 etc
#define PIN_D2 66          // Register 08, pin 2
#define PIN_D3 67          // Register 08, pin 3
#define PIN_D4 68          // Register 08, pin 4
#define PIN_D5 69          // Register 08, pin 5
#define PIN_D6 70          // Register 08, pin 6
#define PIN_D7 71          // Register 08, pin 7

#define PIN_E0 72          // Register 09, pin 0 (9*8)+0=72
#define PIN_E1 73          // Register 09, pin 1 etc
#define PIN_E2 74          // Register 09, pin 2

/////////////////////////////////////////////////// Useful defines

#define FALSE 0            // Logical state 0
#define TRUE 1             // Logical state 1

#define BYTE int           // 8-bit value
#define BOOLEAN short int  // 1-bit value

#define getc getch         // Alternate names..
#define fgetc getch        // ..for identical functions
#define getchar getch
#define putc putchar
#define fputc putchar
#define fgets gets
#define fputs puts

/////////////////////////////////////////////////////////// Control
// Control Functions: RESET_CPU(), SLEEP(), RESTART_CAUSE()
// Constants returned from RESTART_CAUSE() are:
#define WDT_FROM_SLEEP  0   // Watchdog timer has woken MCU from sleep
#define WDT_TIMEOUT     8   // Watchdog timer has caused reset
#define MCLR_FROM_SLEEP 16  // MCU has been woken by reset input
#define NORMAL_POWER_UP 24  // Normal power on reset has occurred

/////////////////////////////////////////////////////////// Timer 0
// Timer 0 (AKA RTCC)Functions: SETUP_COUNTERS() or SETUP_TIMER0(),
//                              SET_TIMER0() or SET_RTCC(),
//                              GET_TIMER0() or GET_RTCC()

// Constants used for SETUP_TIMER0() are:
#define RTCC_INTERNAL    0   // Use instruction clock
#define RTCC_EXT_L_TO_H 32   // Use T0CKI rising edge
#define RTCC_EXT_H_TO_L 48   // Use T0CKI falling edge

#define RTCC_DIV_1       8   // No prescale
#define RTCC_DIV_2       0   // Prescale divide by 2
#define RTCC_DIV_4       1   // Prescale divide by 4
```

```
#define RTCC_DIV_8      2       // Prescale divide by 8
#define RTCC_DIV_16     3       // Prescale divide by 16
#define RTCC_DIV_32     4       // Prescale divide by 32
#define RTCC_DIV_64     5       // Prescale divide by 64
#define RTCC_DIV_128    6       // Prescale divide by 128
#define RTCC_DIV_256    7       // Prescale divide by 256

#define RTCC_8_BIT 0

// Constants used for SETUP_COUNTERS() are the above
// constants for the 1st param and the following for
// the 2nd param:

///////////////////////////////////////////////////////////////// WDT
// Watch Dog Timer Functions: SETUP_WDT() or SETUP_COUNTERS() (see above)
//                      RESTART_WDT()

// Constants used for SETUP_WDT() are:
#define WDT_18MS        8       // Watchdog timer interval=18ms
#define WDT_36MS        9       // Watchdog timer interval=36ms
#define WDT_72MS        10      // Watchdog timer interval=72ms
#define WDT_144MS       11      // Watchdog timer interval=144ms
#define WDT_288MS       12      // Watchdog timer interval=288s
#define WDT_576MS       13      // Watchdog timer interval=576ms
#define WDT_1152MS      14      // Watchdog timer interval=1.15ms
#define WDT_2304MS      15      // Watchdog timer interval=2.30s

//////////////////////////////////////////////////////////// Timer1
// Timer 1 Functions: SETUP_TIMER_1, GET_TIMER1, SET_TIMER1

// Constants used for SETUP_TIMER_1() are:
//      (or (via |) together constants from each group)
#define T1_DISABLED         0       // Switch off Timer 1
#define T1_INTERNAL         0x85    // Use instruction clock
#define T1_EXTERNAL         0x87    // Use T1CKI as clock input
#define T1_EXTERNAL_SYNC    0x83    // Synchronise T1CKI input
#define T1_CLK_OUT          8
#define T1_DIV_BY_1         0       // No prescale
#define T1_DIV_BY_2         0x10    // Prescale divide by 2
#define T1_DIV_BY_4         0x20    // Prescale divide by 4
#define T1_DIV_BY_8         0x30    // Prescale divide by 8

/////////////////////////////////////////////////////////// Timer 2
// Timer 2 Functions: SETUP_TIMER_2, GET_TIMER2, SET_TIMER2
// Constants used for SETUP_TIMER_2() are:
#define T2_DISABLED         0       // No prescale
#define T2_DIV_BY_1         4       // Prescale divide by 2
#define T2_DIV_BY_4         5       // Prescale divide by 4
#define T2_DIV_BY_16        6       // Prescale divide by 16
```

```
///////////////////////////////////////////////////////////////// CCP
// CCP Functions: SETUP_CCPx, SET_PWMx_DUTY
// CCP Variables: CCP_x, CCP_x_LOW, CCP_x_HIGH
// Constants used for SETUP_CCPx() are:
#define CCP_OFF                     0      // Disable CCPx
#define CCP_CAPTURE_FE              4      // Capture on falling edge of
                                          //   CCPx input pin
#define CCP_CAPTURE_RE              5      // Capture on rising edge of
                                          //   CCPx input pi
#define CCP_CAPTURE_DIV_4           6      // Capture every 4 pulses of
                                          //   input
#define CCP_CAPTURE_DIV_16          7      // Capture every 16 pulses of
                                          //   input
#define CCP_COMPARE_SET_ON_MATCH    8      // CCPx output pin goes high
                                          //   when compare succeeds
#define CCP_COMPARE_CLR_ON_MATCH    9      // CCPx output pin goes low
                                          //   when compare succeeds
#define CCP_COMPARE_INT             0xA    // Generate an interrupt when
                                          //   compare succeds
#define CCP_COMPARE_RESET_TIMER     0xB    // Reset timer to zero when
                                          //   compare succeeds
#define CCP_PWM                     0xC    // Enable Pulse Width
                                          //   Modulation mode
#define CCP_PWM_PLUS_1              0x1c
#define CCP_PWM_PLUS_2              0x2c
#define CCP_PWM_PLUS_3              0x3c
long CCP_1;
#byte  CCP_1     =            0x15   // Addresses of CCP1 registers
#byte  CCP_1_LOW =            0x15
#byte  CCP_1_HIGH=           0x16
long CCP_2;
#byte  CCP_2     =            0x1B   // Addresses of CCP2 registers
#byte  CCP_2_LOW =            0x1B
#byte  CCP_2_HIGH=           0x1C

///////////////////////////////////////////////////////////////// PSP
// PSP Functions: SETUP_PSP, PSP_INPUT_FULL(), PSP_OUTPUT_FULL(),
//                PSP_OVERFLOW(), INPUT_D(), OUTPUT_D()
// PSP Variables: PSP_DATA

// Constants used in SETUP_PSP() are:
#define PSP_ENABLED                 0x10   // Enable Parallel Slave Port
#define PSP_DISABLED                0      // Disable Parallel Slave Port

#byte  PSP_DATA=             8      // Address of PSP data register
```

```
/////////////////////////////////////////////////////////////////// SPI
// SPI Functions: SETUP_SPI, SPI_WRITE, SPI_READ, SPI_DATA_IN
// Constants used in SETUP_SSP() are:
#define SPI_MASTER          0x20    // Select SPI master mode
#define SPI_SLAVE           0x24    // Select SPI slave mode
#define SPI_L_TO_H          0       // Strobe data on rising edge of
                                    //   clock
#define SPI_H_TO_L          0x10    // Strobe data on falling edge of
                                    //   clock
#define SPI_DIV_4           0       // Master mode clock divided by 4
#define SPI_CLK_DIV_16      1       // Master mode clock divided by 16
#define SPI_CLK_DIV_64      2       // Master mode clock divided by 64
#define SPI_CLK_T2          3       // Master mode clock source=Timer2/2
#fine SPI_SS_DISABLED       1       // Slave select input disabled

#define SPI_SAMPLE_AT_END   0x8000
#define SPI_XMIT_L_TO_H     0x4000

/////////////////////////////////////////////////////////////////// UART
// Constants used in setup_uart() are:
// FALSE - Turn UART off
// TRUE - Turn UART on
#define UART_ADDRESS        2
#define UART_DATA           4

/////////////////////////////////////////////////////////////////// COMP
// Comparator Variables: C1OUT, C2OUT
// Constants used in setup_comparators() are:    (see 16F877 data
                                                 sheet, figure 12.1)
#define A0_A3_A1_A3              0xfff04  // Two common reference
                                         //   comparators
#define A0_A3_A1_A2_OUT_ON_A4_A5 0xfcf03 // Two independent
                                         //   comparators with outputs
#define A0_A3_A1_A3_OUT_ON_A4_A5 0xbcf05 // Two common reference
                                         //   comparators with outputs
#define NC_NC_NC_NC             0x0ff07  // Comparator inputs
                                         //   disconnected
#define A0_A3_A1_A2             0xfff02  // Two independent
                                         //   comparators
#define A0_A3_NC_NC_OUT_ON_A4   0x9ef01  // One independent
                                         //   comparator with output
#define A0_VR_A1_VR             0x3ff06  // Two comparators with
                                         //   common internal reference
#define A3_VR_A2_VR             0xcff0e  // Two comparators with
                                         //   common internal reference

#bit C1OUT = 0x9c.6
#bit C2OUT = 0x9c.7
```

```
/////////////////////////////////////////////////////////////// VREF
// Constants used in setup_vref() are:
//
#define VREF_LOW        0xa0     // Comparator reference voltage low
                                    range 0-3.75V nominal
#define VREF_HIGH       0x80     // Comparator reference voltage high
                                    range 1.25V-3.75V nominal
// Or (with |) the above with a number 0-15 (reference voltage
   selection within range)
#define VREF_A2         0x40

/////////////////////////////////////////////////////////////// ADC
// ADC Functions: SETUP_ADC(), SETUP_ADC_PORTS() (aka SETUP_PORT_A),
//               SET_ADC_CHANNEL(), READ_ADC()
//
// Constants used for SETUP_ADC() are:    (Fosc=MCU clock frequency)
#define ADC_OFF             0        // ADC Off
#define ADC_CLOCK_DIV_2     0x10000  // ADC clock=Fosc/2
#define ADC_CLOCK_DIV_4     0x4000   // ADC clock=Fosc/4
#define ADC_CLOCK_DIV_8     0x0040   // ADC clock=Fosc/8
#define ADC_CLOCK_DIV_16    0x4040   // ADC clock=Fosc/16
#define ADC_CLOCK_DIV_32    0x0080   // ADC clock=Fosc/32
#define ADC_CLOCK_DIV_64    0x4080   // ADC clock=Fosc/64
#define ADC_CLOCK_INTERNAL  0x00c0   // Internal 2-6us clock

// Constants used in SETUP_ADC_PORTS() are:
#define NO_ANALOGS                                  7    // None - all pins
                                                            are digital I/O
#define ALL_ANALOG                                  0    // A0 A1 A2 A3 A5 E0
                                                            E1 E2 are analog
#define AN0_AN1_AN2_AN4_AN5_AN6_AN7_VSS_VREF        1    // 7 analog, 1
                                                            reference input
#define AN0_AN1_AN2_AN3_AN4                         2    // 5 analog, 3
                                                            digital I/O
#define AN0_AN1_AN2_AN4_VSS_VREF                    3    // 4 analogue, 1
                                                            reference input
#define AN0_AN1_AN3                                 4    // 3 analog, 5
                                                            digital I/O
#define AN0_AN1_VSS_VREF                            5    // 2 analog, 1
                                                            reference input
#define AN0_AN1_AN4_AN5_AN6_AN7_VREF_VREF           0x08 // 6 analog, 2
                                                            reference inputs
#define AN0_AN1_AN2_AN3_AN4_AN5                     0x09 // 6 analog, 2
                                                            digital I/O
#define AN0_AN1_AN2_AN4_AN5_VSS_VREF                0x0A // 5 analog, 1
                                                            reference input
```

```
#define AN0_AN1_AN4_AN5_VREF_VREF    0x0B    // 4 analog, 2 reference
                                                    inputs, 2 digital
#define AN0_AN1_AN4_VREF_VREF        0x0C    // 3 analog, 2 reference
                                                    inputs, 3 digital
#define AN0_AN1_VREF_VREF            0x0D    // 2 analog, 2 reference
                                                    inputs, 4 digital
#define AN0                          0x0E    // 1 analog, 7 digital
#define AN0_VREF_VREF                0x0F    // 1 analog, 2 reference,
                                                    5 digital

// Constants used in READ_ADC() are:
#define ADC_START_AND_READ           7       // This is the default if
                                                    nothing is specified
#define ADC_START_ONLY               1
#define ADC_READ_ONLY                6

///////////////////////////////////////////////////////////// INT
// Interrupt Functions: ENABLE_INTERRUPTS(), DISABLE_INTERRUPTS(),
//                      EXT_INT_EDGE()

// Constants used in EXT_INT_EDGE() are:
#define L_TO_H     0x40 // Interrupt on rising edge of external input
#define H_TO_L     0    // Interrupt on falling edge of external input

// Constants used in ENABLE/DISABLE_INTERRUPTS() are:
#define GLOBAL       0x0BC0 // Identify all interrupts
#define INT_RTCC     0x0B20 // Identify Timer0 overflow interrupt
#define INT_RB       0x0B08 // Identify Port B change interrupt
#define INT_EXT      0x0B10 // Identify RB0 external interrupt
#define INT_AD       0x8C40 // Identify ADC finished interrupt
#define INT_TBE      0x8C10 // Identify RS232 transmit done interrupt
#define INT_RDA      0x8C20 // Identify RS232 receive ready interrupt
#define INT_TIMER1   0x8C01 // Identify Timer1 overflow interrupt
#define INT_TIMER2   0x8C02 // Identify Timer2 overflow interrupt
#define INT_CCP1     0x8C04 // Identify Capture1 or Compare1 interrupt
#define INT_CCP2     0x8D01 // Identify Capture2 or Compare2 interrupt
#define INT_SSP      0x8C08 // Identify Synchronous Serial Port interrupt
#define INT_PSP      0x8C80 // Identify Parallel Slave Port interrupt
#define INT_BUSCOL   0x8D08 // Identify I2C Bus Collision interrupt
#define INT_EEPROM   0x8D10 // Identify EEPROM write completion interrupt
#define INT_TIMER0   0x0B20 // Identify Timer0 overflow interrupt
#define INT_COMP     0x8D40 // Identify Analog Comparator interrupt

#list
#device PIC16F877A
```

Listing 2.21   Disassembled Code for Pulse Output Loop

```
25:  while(1)
26:  {  output_high(PIN_D0);
        006B    1008    BCF    0x8,  0
        006C    1283    BCF    0x3,  0x5
        006D    1408    BSF    0x8,  0
27:     output_low(PIN_D0);
        006E    1683    BSF    0x3,  0x5
        006F    1008    BCF    0x8,  0
        0070    1283    BCF    0x3,  0x5
        0071    1008    BCF    0x8,  0
28:  }
        0072    1683    BSF    0x3,  0x5
        0073    286B    GOTO   0x6b
```

Note the redundancy in the sequence; the pin data direction setting is repeated in each statement, where the file register bank is selected (BCF 0x3,0x5), and the direction bit is cleared to 0 (BCF 0x8,0).

Assembler Block

The maximum output frequency of the pulse waveform can be increased by using a small assembler block to toggle the output bit. A program is shown in Listing 2.22 that outputs a pulse train when a button connected to RB0 input is pressed (active low). The main program provides initialization of the button interrupt and an assembler block, which outputs the signal in a loop that is as short as possible. The interrupt routine at the top of the program is called when the button is not pressed (default condition), switching off the output and waiting for the button to be pressed again to resume the output.

The start of the assembler block is identified by the #asm directive and terminated with #endasm. All the code between these points must conform to the PIC assembler syntax requirements (see Instruction Set, Table 2.11). The interrupt still works, even though it is set up in C, because ultimately the interrupt control settings are the same in C and assembler. Listing 2.23 disassembles the assembly block.

Note that the compiler automatically includes the necessary file register bank select command to access the port data bits. Port B, bit 0, is then set, cleared, and the GOTO takes the execution point straight back to the set instruction, giving a total loop time of

---

Listing 2.22   C Source Code with Assembler Block

```
/*
  Source code file:       FAST.C
  Author, date, version:  MPB 19-10-07 V1.0
  Program function:       Demo of assembler block
  Simulation circuit:     ASSEM.DSN

**********************************************************/

#include "16F877A.h"
#use delay(clock=4000000)

// ISR switches off output and waits for button ************
#int_ext
  void isrext()
  {    output_low(PIN_D0);
       delay_ms(100);
       while(input(PIN_B0));
  }
// Main block initializes interrupt and waits for button ***
void main()
{
  enable_interrupts(int_ext);
  enable_interrupts(global);
  ext_int_edge(L_TO_H);

  // Assembler block outputs high speed pulse wave *******
  #asm

    Start:
      BSF 8,0
      BCF 8,0
      GOTO Start

  #endasm
} // End of source code **********************************
```

four instructions, or 4 μs. The output therefore runs at 250 kHz, 2.5 times faster than the C loop shown in Listing 2.20. If the MCU clock is uprated to the maximum 20 MHz, the output frequency is 1.25 MHz.

A screenshot of this program, FAST.C, under test in MPLAB with VSM debugging is shown in Figure 2.15. The frequency of the output is displayed on the VSM virtual counter/timer instrument.

Table 2.11: PIC 16FXXX Instruction Set by Functional Groups

Operation	Example	
Move		
Move data from F to W	MOVF	0C, W
Move data from W to F	MOVWF	0C
Move literal into W	MOVLW	
Register		
Clear W (reset all bits and value to 0)	CLRW	
Clear F (reset all bits and value to 0)	CLRF	0C
Decrement F (reduce by 1)	DECF	0C
Increment F (increase by 1)	INCF	0C
Swap the upper and lower four bits in F	SWAPF	0C
Complement F value (invert all bits)	COMF	0C
Rotate bits Left through Carry Flag	RLF	0C
Rotate bits Right through Carry Flag	RRF	0C
Clear (reset to 0) the bit specified (e.g., bit 3)	BCF	0C, 3
Set (to 1) the bit specified (e.g., bit 3)	BSF	0C, 3
Arithmetic		
Add W to F	ADDWF	0C
Add F to W	ADDWF	0C, W
Add L to W	ADDLW0F9	
Subtract W from F	SUBWF	0C
Subtract W from F, placing result in W	SUBWF	0C, W
Subtract W from L, placing result in W	SUBLW0F9	
Logic		
AND the bits of W and F, result in F	ANDWF	0C
AND the bits of W and F, result in W	ANDWF	0C, W
AND the bits of L and W, result in W	ANDLW0F9	

Table 2.11: (*Continued*)

Operation	Example
OR the bits of W and F, result in F	IORWF 0C
OR the bits of W and F, result in W	IORWF 0C,W
OR the bits of L and W, result in W	IORLW0F9
Exclusive OR the bits of W and F, result in F	XORWF 0C
Exclusive OR the bits of W and F, result in W	XORWF 0C,W
Exclusive OR the bits of L and W	XORLW0F9
Test and Skip	
Test a bit in F and Skip next instruction if it is Clear (= 0)	BTFSC 0C,3
Test a bit in F and Skip next instruction if it is Set (= 1)	BTFSS 0C,3
Decrement F and Skip next Instruction if it is now 0	DECFSZ 0C
Increment F and Skip next Instruction if it is now 0	INCFSZ 0C
Jump	
Go To a Labeled Line in the Program	GOTO start
Jump to the Label at the start of a Subroutine	CALLdelay
Return at the end of a Subroutine to the next instruction	RETURN
Return at the end of a Subroutine with L in W	RETLW 0F9
Return from Interrupt Service Routine to next instruction	RETFIE
Control	
No Operation, delay for 1 cycle	NOP
Go into Standby Mode to save power	SLEEP
Clear Watchdog Timer to prevent automatic reset	CLRWDT
Load Port Data Direction Register from W*	TRIS06
Load Option Control Register from W	OPTION

Notes: The result of operations can generally be stored in W instead of the file register by adding 'W' to the instruction. General Purpose Register 1, address 0C, represents all file registers (00–4F).
Literal value 0F9 represents all values 00–FF. Bit 3 is used to represent File Register Bits 0–7.
For MOVE instructions data are copied to the destination but retained in the source register.
F = Any file register (specified by number or label), example is 0C.
W = Working register.
L = Literal value (follows instruction), example is 0F9.
* = Use of these instructions not now recommended by manufacturer.

Listing 2.23   Assembler Block Disassembled

```
29: // Assembler block outputs high speed pulse wave *******
30:
31: #asm
32:
33:    Start:
34:    BSF 8,0        006B    1283      BCF  0x3,  0x5
                      006C    1408      BSF  0x8,  0
35:    BCF 8,0        006D    1008      BCF  0x8,  0
36:    GOTO Start     006E    286C      GOTO 0x6c
37:
38: #endasm
```

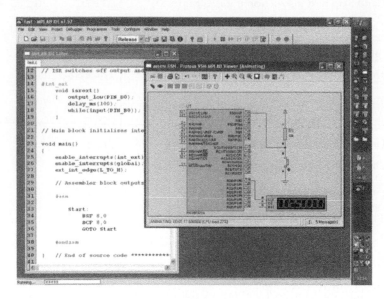

Figure 2.15: Debug Screenshot of FAST.C Showing Output Frequency

PIC Assembly Language

A complete introduction to programming PIC microcontrollers in assembly language is given in *PIC Microcontrollers, An Introduction to Microelectronics* by the author (Elsevier, second edition, 2004). A brief overview is given here for those readers interested primarily in C programming. To program in assembler, some knowledge of the internal hardware of the MCU is needed. The PIC16F877A architecture was introduced in Part 1 of this book, and the file register set is detailed further in Appendix C.

The primary purpose of any programming language is to get data into a system, process it, and output it in some useful form. In assembly language, the program statements act directly on the MCU registers. All the hardware information needed for programming in assembler is given in the data sheet for each PIC MCU, including the instruction set, register details, and setup requirements.

A simplified version of the instruction set is shown in Table 2.11. It is organized by function; that is, instructions with similar functions are grouped together. As explained in Part 1, the operation of the MCU revolves around the numbered file register set and the working register, designated W in the instructions. Register 0C (12_{10}), the first general purpose register, is used to represent the file registers in the examples. The special function registers at the low addresses, which control the MCU setup and program execution, are accessed in exactly the same way as the data registers.

The Move instructions are the most commonly used; these allow a data byte to be moved from the working register to a file register and back or to load immediate data into W. Note that data cannot be moved directly between file registers in the 16FXXX instruction set—this is one of the casualties of the minimal instruction set (RISC) chip design philosophy. The register instructions operate on a single file register, allowing it to be cleared, incremented, decremented, rotated (shifted), and so on. Individual bits may also be set and cleared.

The Arithmetic and Logic instructions operate on pairs of registers in binary: adding, subtracting, and carrying out logical bit-wise operations. If the result of an operation is 0 or a carry or overflow occurs, this is recorded in a flag bit in the status register (SFR 03). For example, if a result is 0, the status register bit 2 is set. The flag can then be used by a bit Test and Skip instruction to select alternate program sequences. In the PIC, this is implemented by the instruction following the test being skipped or not, depending on the result. Usually, this a jump instruction (GOTO or CALL), which takes the program execution point to a new position (or not).

GOTO means go to a given program memory location unconditionally. CALL also means jump but store a return address, so that the current sequence can be resumed when the subroutine is finished, indicated by the RETURN instruction. The jump destination is normally given a label, such as "start" in the example, in the source code.

Unlike C, the program designer must allocate memory explicitly, using suitable labels; variables are declared using an equate directive at the top of the program to identify a GPR for that byte. The register labels are then recognized by the assembler as representing a specific location. Obviously, only 8-bit variables can be used in assembler, so care must be taken if using long values generated in the C program sections. An assembler header file can allocate standard labels to the SFRs in the same way as the C header defines the control register codes. The #include directive is the same in C and assembler and can be used to include assembler header, library, and user source code.

There are only 35 core instructions in the 16FXXX instruction set. This reduced instruction set increases the program execution speed. Additional special instructions are available to compensate for the limited instruction set; these are basically predefined macros. A macro is a code sequence that can be predefined and given its own name, then inserted by the assembler when invoked by name. User-defined macros may also be created as required.

Therefore, if direct control of the MCU registers and instruction sequence is required for any reason or the speed of execution is critical, the C programmer can always revert to assembler code. Since most microcontroller application designers are familiar with assembly language anyway, including assembler blocks typically requires little additional learning time.

Assessment 2
(5 points each, total 100)

1. List the syntax features that a minimal C program must contain if compiled for the PIC16F877A MCU.

2. List the steps required to create and test a C program for a PIC MCU prior to downloading to hardware.

3. Write a C statement that outputs the 8-bit value 64_{10} to Port C. Write an alternative 1-bit output statement that has the same effect, assuming all the port bits are initially 0.

4. Describe briefly the difference between a WHILE loop, a DO..WHILE loop, and a FOR loop.

5. Describe the effect of the following statements on active high LEDs connected to Port D, assuming an active low switch circuit is connected to pin RC7:

```
output_D(255); delay_ms(1000);
while(!input(PIN_C7)){output_D(15);}
output_D(0);
```

6. Calculate the highest positive number that can be represented by the following variable types: (a) 8-bit unsigned integer, (b) 16-bit signed integer, (c) 32-bit floating point number.

7. Estimate the degree of precision provided by the following numerical types as a percentage, to two significant figures: (a) 8-bit integer, (b) 32-bit FP number.

8. Work out the value of the FP number represented by the binary code

   ```
   1000 0010 0011 0000 0000 0000 0000 0000
   ```

9. Write a C statement to convert numbers 0 to 9 to their ASCII hex code, using variables 'n' for the number and 'a' for the ASCII code and send it to serial LCD.

10. State the result of each of these operations in decimal and 4-bit binary, if n = 5 and m = 7:

    (a) `n++`.
    (b) `~m`.
    (c) `n&m`.
    (d) `n|m`.
    (e) `n^m`.

11. State the effect of the jump commands `continue`, `break`, and `goto label` when used within a program loop.

12. A menu is required with a choice of three options to be selected by a numerical variable x = 1, 2, 3. Each option is implemented in a separate function, `funx()`. Write a C code section to show how `switch` can be used to implement the menu.

13. Explain why the use of local variables is preferable in C programs designed for microcontrollers with limited RAM.

14. Explain how the use of functions leads to well-structured C programs and the benefits of this design approach.

15. State the meaning of the source code items that are underlined:

```
int out(int16 t)

  {
    int16 n;

  while (input(PIN_D0))
    {  outbyte++;
      for(n=1;n<t;n++);
    }
    return outbyte;
}
```

16. Outline briefly the format of the RS232 signal and how it is used to operate a serial alphanumeric LCD.

17. Draw a simple flowchart to represent a function to scan the keys of a numerical keypad and return a code for a key press.

18. Explain the meaning of each component of the statement
`printf("%d",anum[n]);`

19. Explain the significance of the & and * operators in C.

20. State the function of the compiler directives:

    (a) `#include`.
    (b) `#define`.
    (c) `#use`.
    (d) `#device`.
    (e) `#asm`.

Assignments 2

To undertake these assignments, install Microchip MPLAB (www.microchip.com), Labcenter ISIS Lite (www.proteuslite.com), and CCS C Lite (www.ccsinfo.com). Application files may be downloaded from www.picmicros.org.uk. Run the applications in MPLAB with Proteus VSM selected as the debug tool. Display the animated schematic in VSM viewer, with the application COF file attached to the MCU (see the appendices for details).

Assignment 2.1

Download the OUTBYTE.DSN file and attach ENDLESS.COF. Check that it works correctly. Modify the program so that the LED output LSB flashes at 4 Hz. Predict the frequency of the MSB and measure it using the simulation clock.

Assignment 2.2

Download the SIREN project files and check that the SIREN program in Listing 2.7 works correctly. Modify the program to produce a default output at 1 kHz. Further modify the program so that the output frequency is halved each time the input button is pressed.

Assignment 2.3

Download the CALC project files and check that the CALC program works correctly.
Modify the program such that the ON/C key must the pressed to start the program
and pressing it again disables the program. Investigate the use of the string processing
functions to provide a more elegant implementation of the conversion of an input string of
numbers to decimal during the input phase. Outline how the program could be developed
to handle floating point numbers to provide a more practical calculator.

Assignment 2.3

Download the CALC project files and check that the CALC C program works correctly. Modify the program such that the ONCE key must be pressed to start the program and pressing it again disables the program. Investigate the use of the string processing functions to provide a more elegant implementation of the conversion of an input string of numbers to decimal during the input phase. Outline how the program could be developed to handle floating point numbers to provide a more practical calculator.

PART 3

C Peripheral Interfaces

3.1 PIC16 C Analog Input

- Analog input display

- Voltage measurement

- ADC setup codes

A microcontroller analog input allows an external voltage to be converted to digital form, stored, and processed. This type of input occurs in data loggers, control systems, digital audio, and signal processors, to mention just a few. The dsPIC range is designed specifically for high-speed analog signal processing.

Analog Setup

A basic setup to demonstrate analog input is shown in Figure 3.1. The PIC16F877 has eight analog inputs, which are accessed via RA0, RA1, RA2, RA3, RA5, RE0, RE1, and RE2, being renamed AN0 to AN7 in this mode. All these pins default to analog operation, but a combination of analog and digital inputs can be selected using the system function set_up_adc_ports().

These inputs are multiplexed into a single converter, so they can be read only one at a time. The function set_ADC_channel(n) selects the input channel. The analog-to-digital converter module has a resolution of 10 bits, giving a binary output of 0×000 to $0 \times 3FF$ (1023_{10}). Therefore, the measurement has a precision of $1/1024 \times 100\%$, which is slightly better than 0.1%. This is good enough for most practical purposes. A 16-bit integer or floating point variable is needed to receive this result.

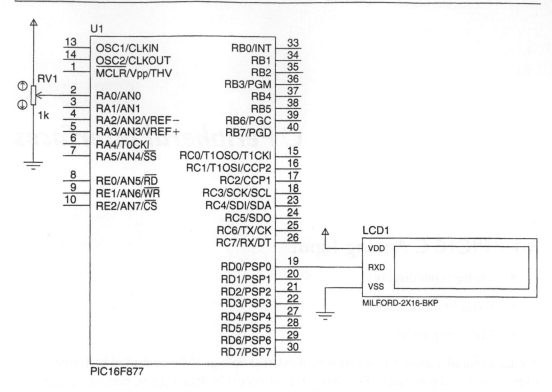

Figure 3.1: Single Analog Input and Display Test Circuit

Alternatively, the low-resolution mode can be used if an 8-bit conversion is sufficiently precise (output = 0–255). This mode is selected using the directive #device ADC=8. The function read_ADC() then returns the input value as an unsigned integer. The default input voltage range is 0–5 V, which does not give an exact conversion factor. In the demo program, Listing 3.1, the 8-bit input value is divided by 32 to give an arbitrary voltage level from 0 to 8. This is then converted to the ASCII code by adding 0x30 and sending it to the display. The operation is repeated endlessly, using the statement for(;;), which means execute a for loop unconditionally.

Voltage Measurement

The circuit shown in Figure 3.2 allows the input voltage at each analog input to be displayed. An external reference voltage (2.56 V) is connected to RA3, which sets the maximum of the input range. This allows a more accurate and convenient scaling of the measurement. The reference voltage is supplied by a zener diode and voltage divider

Listing 3.1   Source Code for Simple Analog Input Test Program

```
/*  ANALIN.C MPB 5-1-07
    Read & display analog input

*************************************************************/

#include "16F877A.h"
#device ADC=8                                   //8-bit conversion

#use delay(clock=4000000)
#use rs232(baud=9600, xmit=PIN_D0, rcv=PIN_D1)  //LCD output

void main() //***********************************************
{
  int vin0;                                     // Input variable

  setup_adc(ADC_CLOCK_INTERNAL);                // ADC clock
  setup_adc_ports(ALL_ANALOG);                  // Input combination
  set_adc_channel(0);                           // Select RA0

  for(;;)
  {   delay_ms(500);
      vin0=read_adc();                          //Get input byte
      vin0=(vin0/32)+0x30;                      //Convert to ASCII

      putc(254); putc(1); delay_ms(10);         // Clear screen
      printf("Input="); putc(vin0);             // Display input
  }
}
```

circuit. The value of the zener load resistor has been selected by simulation to adjust the voltage to $2.560 \pm 0.1\%$. A potentiometer is connected to each of the measured inputs so it can be set to an arbitrary test value. The test program VOLTS.C is provided in Listing 3.2.

This time, the ADC resolution is set to 10 bits, to obtain a more precise reading. Floating point array variables are declared for the input readings (0–1023) and the calculated voltage. The reference voltage, 2.56 V, is represented by the maximum conversion value, 1024, so the scaling factor is $1024/2.56 = 400$ bits per volt. The input is therefore divided by this factor to obtain a display in volts. Note that, in the division operation, both values must be float types.

The ADC port setup code selects all inputs as analog, with RA3 an external reference (although this is not obvious from the select statement format). All the possible

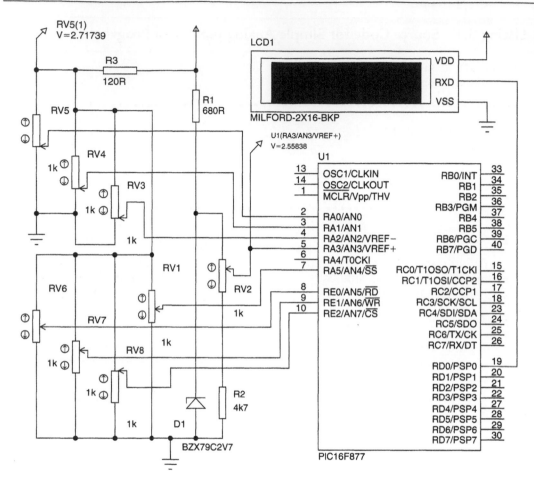

Figure 3.2: Input Voltage Measurement and Display

combinations of analog and digital inputs are given in the 16F877A.H header file, Listing 2.19. When the program is compiled, the define statement selected is replaced by the corresponding hex code, which is then loaded into the ADC control register to set up the ADC.

The set of functions that control the ADC are listed in Table 3.1. The function `setup_adc()` allows the clock rate (ADC sampling rate) to be selected to suit the application, and `setup_adc_ports()` allows the mix of analog and digital inputs to be defined using the combinations provided in the header file.

Listing 3.2   Test Program for Voltage Measurement

```c
/* VOLTS.C  MPB  25-3-07
   Read & display 10-bit input voltage

***********************************************************/

#include "16F877A.h"
#device ADC=10                                  // 10-bit operation
#use delay(clock=4000000)
#use rs232(baud=9600,xmit=PIN_D0,rcv=PIN_D1)

void main()  //***********************************************
{
  int  chan;
  float     analin[8], disvolts[8];             // Array variables

  setup_adc(ADC_CLOCK_INTERNAL);                // ADC Clock source

  setup_adc_ports(AN0_AN1_AN2_AN4_AN5_AN6_AN7_VSS_VREF);  // ADC inputs

    while(1)                                     // Loop always
    {
      for(chan=0;chan<8;chan++)                  // Read 8 inputs
      { delay_ms(1000);                          // Wait 1 sec
        set_adc_channel(chan);                   // Select channel
        analin[chan]=read_adc();                 // Get input
        disvolts[chan]=(analin[chan])/400;       // Scale input
        putc(254);putc(1);delay_ms(10);          // Clear display
        printf(" RA%d = %4.3g",chan,disvolts[chan]);  // Display volts
      }
    }
}
```

Table 3.1: CCS C Analog Input Functions

Action	Description	Example
ADC SETUP	Initialize ADC	setup_adc(ADC_CLOCK_INTERNAL);
ADC PINS SETUP	Initialize ADC pins	setup_adc_ports(RA0_ANALOG);
ADC CHANNEL SELECT	Select ADC input	set_adc_channel(0);
ADC READ	Read analog input	inval=read_adc();

Listing 3.3   External Interrupt Test Program Source Code

```
// INTEXT.C MPB 10-4-07
// Demo external interrupt RB0 low interrupts foreground output count

#include "16F877A.h"
#use delay(clock=4000000)

#int_ext                                // Interrupt name
void isrext()                           // Interrupt service routine
  { output_D(255);                      // ISR action
    delay_ms(1000);
  }

void main()   //**********************************************
{
  int x;

  enable_interrupts(int_ext);           // Enable named interrupt
  enable_interrupts(global);            // Enable all interrupts
  ext_int_edge(H_TO_L);                 // Interrupt signal polarity

  while(1)                              // Foreground loop
  {
    output_D(x); x++;
    delay_ms(100);
  }
}
```

Interrupt Example

Program INTEXT.C (Listing 3.3) demonstrates the basic interrupt setup. An output count represents the primary task. This is interrupted by the switch input at RB0 going low, forcing the execution of the interrupt service routine, which causes all the output LEDs to come on for 1 second. The original task is then automatically resumed at the point where it was interrupted. It is designed to run on the hardware shown in schematic Figure 3.3.

When the RB0 interrupt is detected during the main loop, the context (current register contents) is saved before the ISR executed. If the program execution is studied carefully, it can be seen that the original count prior to the interrupt is restored to the port output after the interrupt. The ISR includes code to save and restore the MCU registers, so that the main task can be resumed unaffected by the interrupt. Only local variables should be used in the ISR to protect the integrity of the rest of the program.

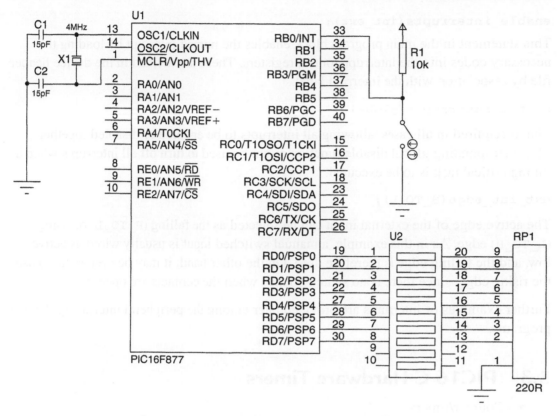

Figure 3.3: External Interrupt Test Hardware

Interrupt Statements

The program statements associated with interrupt operation are as follows.

#int_ext

This directive tells the compiler that the code immediately following is the service routine for this particular interrupt. The routine is the form in a standard function, with a function name appropriate to the ISR task, in this case void isrext(). The interrupt name is preceded by # (hash) to mark the start of the ISR definition and to differentiate it from a standard function block. An interrupt name is defined for each interrupt source.

```
enable_interrupts(int_ext);
```

This statement in the main program block enables the named interrupt by loading the necessary codes into the interrupt control registers. These are defined in the device header file by association with the interrupt label.

```
enable_interrupts(global);
```

This is required in all cases, allowing all interrupts to be enabled or disabled together. The corresponding global disable function might be used to turn off all interrupts when a timing critical task is to be executed.

```
ext_int_edge(H_TO_L);
```

The active edge of the external input can be selected as the falling (H_TO_L) or rising (L_TO_H) edge. As in this example, a manual switched input is usually wired as active low, and the falling edge is therefore used. On the other hand, it may be preferable to use the rising edge, since there is no switch bounce when the contacts are opening.

Further examples of interrupts are provided later among the peripheral interfacing demo programs.

3.3   PIC16 C Hardware Timers

- Counter/timers

- Capture and Compare

- Timer interrupt

The PIC 16F877 has three hardware timers built in: Timer0 (originally called RTCC, the real-time counter clock), Timer1, and Timer2. The principal mode of operation of these registers are as counters for external events or timers using the internal clock. Additional registers are used to provide Capture, Compare, and Pulse Width Modulation (PWM) modes. The CCS timer function set is shown in Table 3.5.

Counter/Timer Operation

A counter/timer register consists of a set of bistable stages (flip-flops) connected in cascade (8, 16, or 32 bits). When used as a counter, a pulse train fed to its least significant bit (LSB) causes the output of that stage to toggle at half the input frequency. This is fed to the next significant bit, which toggles at half that rate, and so on. An 8-bit counter thus

Table 3.5: Timer Functions

Action	Description	Example
TIMERX SETUP	Set up the timer mode	setup_timer0(RTCC_ INTERNAL|RTCC_DIV_8);
TIMERX READ	Read a timer register (8 or 16 bits)	count0 = get_timer0();
TIMERX WRITE	Preload a timer register (8 or 16 bits)	set_timer0(126);
CCPX SETUP	Select PWM, capture, or compare mode	setup_ccp1(ccp_pwm);
PWMX DUTY	Set PWM duty cycle	set_pwm1_duty(512);

counts up from 0x00 to 0xFF (255) before rolling over to 0 again (overflow). The binary count records the number of clock pulses input at the LSB.

In the '877, Timer0 is an 8-bit register that can count pulses at RA4; for this purpose, the input is called T0CKI (Timer0 clock input). Timer1 is a 16-bit register that can count up to 0xFFFF (65,535) connected to RC0 (T1CKI). The count can be recorded at any chosen point in time; alternatively, an interrupt can be generated on overflow to notify the processor that the maximum count has been exceeded. If the register is preloaded with a suitable value, the interrupt occurs after a known count.

The counters are more frequently used as timers, with the input derived from the MCU clock oscillator. Since the clock period is accurately known, the count represents an accurate timed period. It can therefore be used to measure the period or frequency of an input signal or internal intervals or generate a regular interrupt. Many PIC MCUs incorporate one or more Capture, Compare, and PWM (CCP) modules that use the timer registers.

A timer/counter register may have a prescaler, which divides the input frequency by a factor of 2, 4, 8, and so forth using additional stages, or a postscaler, which does the same at the output. Timer0 has a prescaler that divides by up to 128; Timer1 has one that divides by 2, 4, or 8; and Timer2 has a prescaler and postscaler that divide by up to 16.

PWM Mode

In Pulse Width Modulation mode, a CCP module can be used to generate a timed output signal. This provides an output pulse waveform with an adjustable high (mark) period.

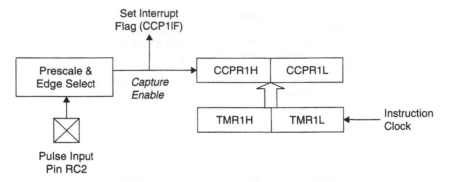

Figure 3.5: Capture Hardware Block Diagram

Listing 3.5 Capture Mode Demo Program

```
// PERIOD.C MPB 11-4-07
// Demo of period measurement

#include "16F877A.h"  //***************************

#int_ccp1                        // Interrupt name
  void isr_ccp1()                // Interrupt function
  {
    set_timer1(0);               // Clear Timer1
    clear_interrupt(INT_CCP1);   // Clear interrupt flag
  }

void main()  //***********************************
{
  setup_timer_1(T1_INTERNAL);    // Internal clock
  setup_ccp1(CCP_CAPTURE_RE);    // Capture rising edge on RC2

  enable_interrupts(GLOBAL);     // Enable all interrupts
  enable_interrupts(INT_CCP1);   // Enable CCP1 interrupt
  while(1){}
}
```

capture event. The captured value is copied automatically into a variable called CCP_1. The simulation of this program is shown in Figure 3.6. When the program is run with the 100-Hz signal input, a count of 9963 μs is captured (error = 0.4%). This shows that some allowance may be needed for the software overhead associated with the capture process and adjustment made to correct the result obtained.

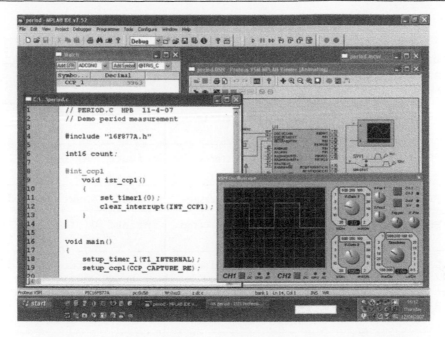

Figure 3.6: Capture Mode Used to Measure Input Period

3.4   PIC16 C UART Serial Link

- RS232 port functions

- Simulation with virtual terminal

A basic serial link is provided by the UART. We have already seen that any pair of pins can be used for this interface, as the data rate is quite low, allowing the signals to be generated in software. However, a dedicated hardware port is provided, which must be used if an interrupt is needed. The CCS C library functions associated with this port are listed in Table 3.6.

The UART can be tested in simulation mode by connecting it to the virtual terminal provided in Proteus VSM, as shown in Figure 3.7. The terminal input RXD (receive data) is connected to the PIC MCU TX (transmit) pin, and the TXD (transmit data) output is connected to PIC MCU RX (receive). It has additional handshaking (transmission control) lines RTS and CTS, but these are not usually needed.

Table 3.6: RS232 Serial Port Functions

Title	Description	Example
RS232 SET BAUD RATE	Set hardware RS232 port baud rate	`setup_uart(19200);`
RS232 SEND BYTE	Write a character to the default port	`putc(65)`
RS232 SEND SELECTED	Write a character to selected port	`s=fputc("A",01);`
RS232 PRINT SERIAL	Write a mixed message	`printf("Answer:%4.3d",n);`
RS232 PRINT SELECTED	Write string to selected serial port	`fprintf(01,"Message");`
RS232 PRINT STRING	Print a string and write it to array	`sprintf(astr,"Ans=%d",n);`
RS232 RECEIVE BYTE	Read a character to an integer	`n=getc();`
RS232 RECEIVE STRING	Read an input string to character array	`gets(spoint);`
RS232 RECEIVE SELECTED	Read an input string to character array	`astring=fgets(spoint,01);`
RS232 CHECK SERIAL	Check for serial input activity	`s=kbhit();`
RS232 PRINT ERROR	Write programmed error message	`assert(a<3);`

The program listed as HARDRS232.C (Listing 3.6) is attached to the MCU in the simulator. The `getc()` function is used to read a character from the virtual terminal; it waits for user input. The terminal must be activated by clicking inside terminal window, and the computer keyboard then provides the input to the PIC as the corresponding ASCII codes; these are assigned to the variable `incode`, as they arrive.

The ASCII code can be output using `printf()`. If formatted as a decimal, the numerical value of the character code is displayed. Alternatively, the character formatting code `%c` is used to display the character itself. The function `putc(13)` outputs the code for a line return on the display. If `putc()` is used to output an ASCII code, the character is displayed.

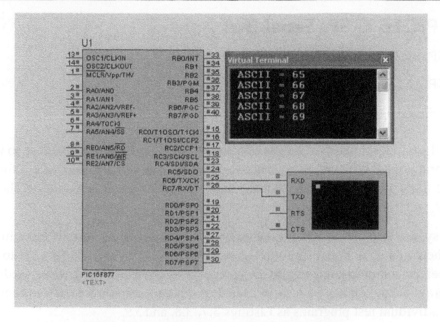

Figure 3.7: RS232 Peripheral Simulation

Listing 3.6 Hardware UART Demo Program

```
// HARDRS232.C MPB 13-6-07
// Serial I/O using hardware RS232 port

#include "16F877A.h"
#use delay(clock=8000000)          // Delay function needed for RS232
#use rs232(UART1)                  // Select hardware UART

void main()  //************************************
{
  int incode;
  setup_uart(9600);                      // Set baud rate

  while(1)
  { incode = getc();                 // Read character from UART
    printf(" ASCII = %d ",incode);   // Display it on
    putc(13);                                // New line on display
  }
}
```

3.5   PIC16 C SPI Serial Bus

- SPI system connections

- SPI function set

- SPI test system

The serial peripheral interface master controller uses hardware slave selection to identify a peripheral device with which it wishes to exchange data (refer to Section 1.4 for full details of the signaling protocol). The available set of SPI driver functions are shown in Table 3.7.

The test system has a slave transmitter that reads a binary-coded decimal input from a thumbwheel switch and sends it to the master controller. This resends the code to the slave receiver, which outputs to a BCD display (0–9). Each of three devices needs its own test program to make the system work. The test system hardware is shown in Figure 3.8 and the individual test programs as Listings 3.7, 3.8, and 3.9.

As seen in the schematic, the slave MCUs are permanently enabled by connecting their slave select inputs to ground. This is possible because there is only one sender on the master input, so there is no potential contention. In a system with more that one slave sender, each would need a separate slave select line, with only one being enabled at a time.

The individual programs were created as separate projects in MPLAB but saved in the same folder, sharing a copy of the MCU header file. The COF files were then attached to the corresponding chip in the simulated hardware.

Table 3.7: SPI Function Set

Operation	Description	Example
SPI SETUP	Initializes SPI serial port	setup_spi(spi_master);
SPI READ	Receives data byte from SPI port	inbyte=spi_read();
SPI WRITE	Sends data byte via SPI port	spi_write(outbyte);
SPI TRANSFER	Sends and receives via SPI	inbyte=spi_xfer(outbyte);
SPI RECEIVED	Checks if SPI data received	done=spi_data_is_in();

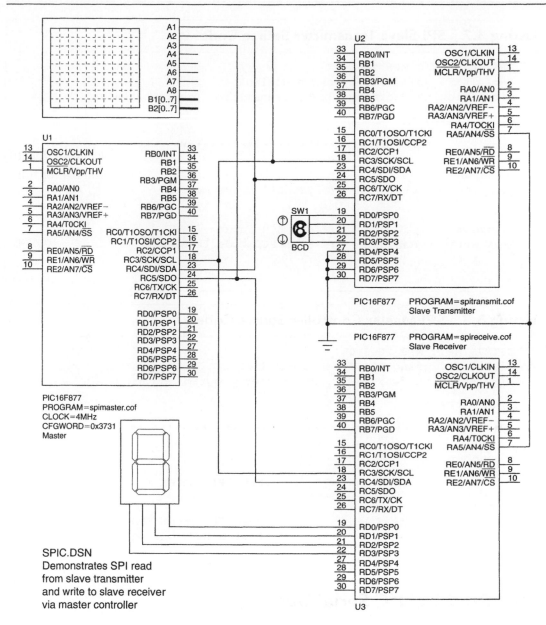

Figure 3.8: SPI Test System Schematic

Listing 3.7   SPI Slave Transmitter Source Code

```
// SPITRANSMIT.C MPB 20-6-07
// Serial I/O using SPI synchronous link
// Simulation hardware SPIC.DSN, transmitter program attached to U2

#include "16F877A.h"

void main()   //******************************************
{
  int sendnum;
  setup_spi(spi_slave);      // Set SPI slave mode

while(1)
  { sendnum = input_D();     // Get BCD input
    spi_write(sendnum);      // Send BCD code to master
  }
}
```

Listing 3.8   SPI Master Controller Source Code

```
// SPIMASTER.C MPB 20-6-07
// Serial I/O using SPI synchronous link
// Simulation hardware SPIC.DSN, master program, attach to U1

#include "16F877A.h"

void main()   //**********************************************************
{
  int number;

  setup_spi(spi_master);                // Set SPI master mode

  while(1)
  { number = spi_read();                // Read SPI input BCD code
    spi_write(number);                  // Resend BCD code to slave
  }
}
```

3.6   PIC16 C I²C Serial Bus

- I²C simulation test system

- I²C control, address, and data bytes

The inter-integrated circuit (I²C) synchronous serial bus provides a means of exchanging data between peripheral devices and microcontrollers using software

Listing 3.9 SPI Slave Receiver Source Code

```
// SPIRECEIVE.C MPB 20-6-07
// Serial I/O using SPI synchronous link
// Simulation hardware SPI.DSN, receiver program, attach to U3

#include "16F877A.h"

void main()  //****************************************
{
  int recnum;
  setup_spi(spi_slave);      // Set SPI slave mode

  while(1)
  { recnum=spi_read();       // Read BCD code at SPI port
    output_D(recnum);        // Display it
  }
}
```

addressing. This means that only two signals are required, data and clock (see Section 1.4 for details).

The test system shown in Figure 3.9 has only one I^2C peripheral device, the 24AA256 serial flash memory chip, to keep it as simple as possible. Serial memory is a common feature of applications that require additional data storage, such as a data logger. It allows the internal EEPROM of the PIC to be expanded using only two I/O pins. The downside is that the memory access is rather slow, with the maximum write cycle time of 5 ms (200 bytes/sec) specified for this device. Therefore, the data sampling rate needs to be suitably modest.

The serial memory chip has a capacity of 256-k bits, or 32-k bytes, with three external address pins: A0, A1, and A2. This allows a set of up to eight chips to be used in the system, each with a different hardware address, 0–7. This address is included in the address code sent by the master controller, so that a specific address in a selected chip can be accessed. With eight 32-k chips, the total address space is 256 k. In the test system, the memory chip hardware address is 000.

The system reads a test code set manually on Port B inputs, which is copied to the serial memory. Pull-ups must be fitted to the serial clock and data lines, and a virtual I^2C analyzer is also attached to the bus. The test program writes the test byte (3F in the

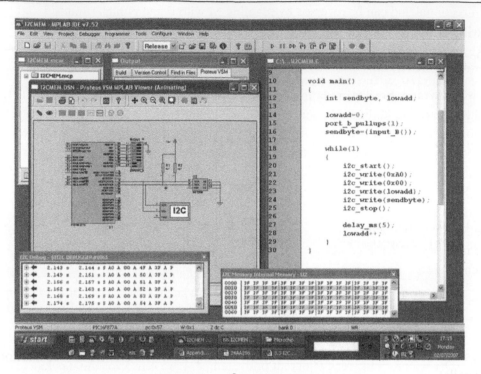

Figure 3.9: I²C Test System

example shown) to the address `lowadd`, which increments from 0 after each write. The `i2c_start()` function initiates the data transfer sequence, by generating a start bit on the data line. This is followed by 4 bytes, containing control, address, and data codes.

The first is the control code, A0. The memory chip has a factory-set high address code of 0101(A). This distinguishes it from other types of I²C devices that may be added to the bus. The next 3 bits are the hardware address (000), and the LSB is set to 0 to indicate a write operation, making the low nibble 0000. This is followed by the two address bytes. The high address byte is 00, and the low address increments from 0, so the test program writes only to the first 256 bytes. The data byte follows, which is read in from the input switches.

Each of these bytes must be acknowledged by the receiving device taking the data line low, and the transfer is terminated by a stop bit. More details on the exact data format and timing requirements may be found in the chip data sheet.

The simulation system allows the bus activity to be logged and displayed in the I²C debug window using the virtual bus monitor instrument. A time stamp, the transfer codes, and

Table 3.8: I²C Functions

Operation	Description	Example
I2C WRITE	Send a single byte	`i2c_write(outbyte);`
I2C READ	Read a received byte	`inbyte=i2c_read();`
I2C STOP	Issue a stop command in master mode	`i2c_stop();`
I2C POLL	Check to see if byte received	`sbit=i2c_poll();`

the Start (S), Acknowledge (A), and Stop (P) bits are detected as they occur. In addition, the memory contents can be displayed to confirm the test data and which locations have been written.

When the memory content window is opened, we see that it retains the data from previous runs of the simulation, representing the nonvolatile nature of the data store. To see the data change, a new code must be set on the switches for each run.

The I²C functions are summarized in Table 3.8.

3.7   PIC16 C Parallel and Serial Interfaces

- PSP functions and test system
- Comparison of parallel and serial links

The parallel slave port (PSP) allows an external controller to initiate an 8-bit data exchange with the PIC MCU. This method of data exchange is compared with the serial ports.

Parallel Slave Port

In the example in Figure 3.10, a master '877 is feeding data to a slave chip of the same type. Arbitrary data are set on the DIP switch at Port B of the master. The internal pull-ups available on these pins are activated in the master program to avoid the need for external resistors on the switches. The test data are transferred to Port C and presented to the slave Port C pins (Listing 3.10). The slave port is already enabled via E0 (!CS = not

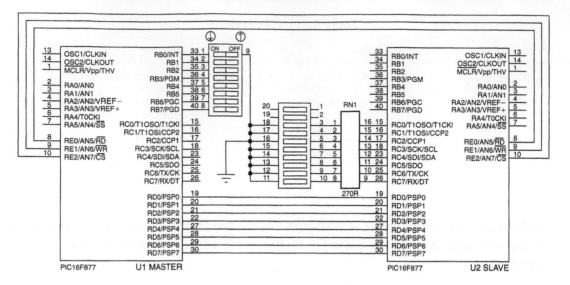

Figure 3.10: PSP Test System

Listing 3.10   PSP Master Test Program

```
// PSPMASTER.C
// Test system master controller program, design file PSP.DSN, U1

#include "16F877A.h"

void main()  //****************************************
{
  int sendbyte;
  port_b_pullups(1);              // Activate Port B pull-ups

  while(1)
  { sendbyte = input_B();         // Get test byte
    output_D(sendbyte);           // Output on PSP bus

    output_low(PIN_E2);           // Select PSP slave
    output_low(PIN_E1);           // Write byte to slave port
    output_high(PIN_E1);          // Reset write enable
  }
}
```

Listing 3.11   PSP Slave Test Program

```
// PSPSLAVE.C
// Test system slave controller program, design file PSP.DSN, U2

#include "16F877A.h"

void main()   //*****************************************
{
  int recbyte;
  setup_psp(PSP_ENABLED);          // Enable PSP slave port

  while(1)
  { if(psp_input_full())           // If data have been received
    { recbyte=input_D();           // Copy in test data
      output_C(recbyte);           // Display data on bar graph
    }
  }
}
```

---

chip select) on Port E, and the data are latched in when E1 (!WR = not write) is pulsed low by the master.

In simulation mode, the write pulse frequency was measured at 40 kHz (MCU clock = 4 MHz). The slave program (Listing 3.11) monitors the receive flag associated with the port and picks up the data when the port indicates that data have been loaded into the PSP data register. The data then are transferred to Port C for display on the bar graph.

A parallel external bus can thus be created that connects microcontrollers, extra memory, and other 8-bit devices to form a system similar to a conventional microprocessor system. On the PSP bus, the master must select the peripheral device to be accessed using the chip select mechanism. If necessary, an address decoding system can be added to expand the hardware without using extra master pins. For example, a 3-bit decoder generates eight chip select signals. A memory space is created for the master, where different peripherals are accessed at separate address ranges.

Table 3.9 summarizes the PSP functions.

Comparison of Communication Links

We can now compare the available PIC MCU communication ports so that the most suitable can be selected for any given application. Table 3.10 summarizes the main features.

Table 3.9: PSP Functions

Operation	Description	Example
PSP SETUP	Enables or disables PSP	setup_psp(PSP_ENABLED);
PSP DIRECTION	Sets the PSP data direction	set_tris_e(0);
PSP OUTPUT READY	Checks if output byte is ready to go	pspo = psp_output_full();
PSP INPUT READY	Checks if input byte is ready to read	pspi = psp_input_full();
PSP OVERFLOW	Checks for data overwrite error	pspv = psp_overflow();

As we have seen, three serial communication interfaces are available plus the parallel slave port.

In theory, the parallel port should be the fastest, because 8 bits can be transferred at a time. The PSP can be used to create a multiprocessor system with a common data bus connected to same port on other MCUs, with one master controlling the addressing system and selecting the slave MCU. One example of such a multiprocessor system is a robot with a separate controller for each motor. The master controller sends data to the motor slaves to set position, speed, or acceleration of that axis. Data transfer speed may be crucial to optimum system performance, so the parallel connection may be preferred in this case. This is feasible as long as the physical distance between the controller and the motors is not too far.

For serial data transfer, speed (bits per second) increases as we progress from UART through I^{2}C to SPI. As well as being the fastest, SPI is also relatively simple to implement. It can operate in Multimaster mode but needs hardware slave selection. I^{2}C needs only two wires and operates like a mini-network, so it may be more effective for larger systems. However, the software is more complex and carries a significant addressing overhead. The UART is a simple way to link a single master and slave and allows greater link distance by use of line drivers. On the other hand, it does not support any form of multiprocessor or bus system.

Table 3.10: Comparison of PIC Communication Ports

	UART	SPI	I²C	PSP
Description	Serial RS232, Host-terminal, single link	Serial data, bus connection with hardware selection	Serial data and address, bus connection with software addressing	Parallel 8-bits, bus connection with hardware control
Clock	Asynchronous	Synchronous, max 5 MHz	Synchronous, max 5 MHz	Synchronous
Wiring	TX, RX, GND	SCK, SDI, SS	SCL, SDA + 10-k pull-ups	PSP0–PSP7, RD, WR, CS
Data	6–9 bits	8 bits serial	8 bits + address + control Page mode option	8 bits parallel
Control	Start, Stop bits	Clock strobe	Clock strobe, Start, Acknowledge	Read, Write, Chip Select
Speed (bits/sec)	LOW <19.2 kb/sec	HIGH <5 Mb/sec	HIGH <1–5 Mb/sec, depends on mode	MID <40 × 8 = 240 kb/sec[1]
Distance[2]	HIGH <100 m	LOW <1 m	LOW <1 m	LOW <1 m
Nodes	2 only	Unlimited[3]	1024 (10-bit address)	Limited by bus characteristics
Systems	Single peer to peer	Master/slave	Master/slave	Master/slave
Operation	Can be connected as a simple 2-wire system but has additional handshaking modes and parity checking for extra reliability	Simple clocked data, high speed but requires slave selection wiring and possibly external decoding	Complex software control and addressing reduces speed but requires no slave selection wiring or external decoding hardware	Simple hardware control but with limited bus length. Higher speeds possible using assembler routine. May need external decoding.
Typical applications	PC host to MCU target data transfer (e.g., data logger)	Sensor data link, MCU communication link	Multiperipheral control system with sensors and low-speed memory data storage	Multiprocessor system, parallel MCU data link

Notes:
[1] This is an estimated speed using nonoptimized C code to drive the bus. If optimized assembler code were used, this could be improved significantly.

[2] Transmission distance in the UART is enhanced by using line drivers to increase the signal voltage to overcome line impedance and interference. Data transmission at TTL signal levels in the other links restricts the distance to within the same subsystem (board, unit, or back plane). For greater distances and multinode operation, a local area network interface is required, which provides synchronous data communication with unlimited software addressing and error correction.

[3] The SPI system can be expanded by additional address decoding and line drivers as necessary, but there are practical limits to this option, and I²C or networking would probably be more effective.

3.8   PIC16 C EEPROM Interface

- EEPROM test system

- EEPROM test program

The internal electrically erasable programmable read only memory block is not strictly speaking a peripheral, as it is internal to the MCU, but it is accessed in a way similar to external devices so it is included in this part. In the 16F877, the EEPROM is a block of 256 bytes of nonvolatile read/write memory. It allows data to be retained while the power is off, which is useful in applications such as an electronic lock where a secure code needs to be stored.

Figure 3.11 shows a test circuit that demonstrates its operation. Arbitrary 8-bit codes are set on the switch bank, which are stored, recalled, and displayed on the LED bank. The R/!W (Read/Not Write) input switch is closed to select the Write mode. The switch code is set and the button pressed. This stores the code in the first EEPROM location,

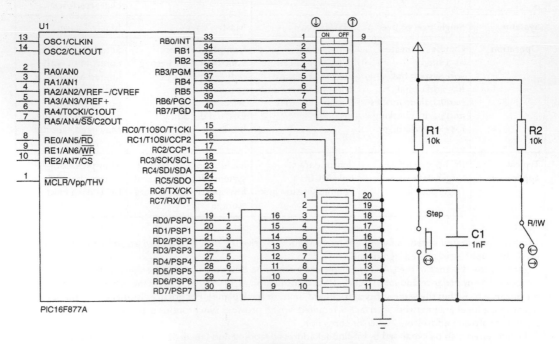

Figure 3.11: EEPROM Test System

address 0. The switch code is then changed and the next code stored in location 1, and so on until a 0 is entered on the switches. As the data are stored, each byte is displayed on the bar graph.

The R/!W switch is then opened to select read mode. As the button is pressed, the same sequence of stored codes is displayed from memory. The nonvolatile data storage is demonstrated by the fact that the test data are retained between successive simulation runs. This can be viewed if the simulation is paused and the EEPROM data window selected from the debug menu. Listing 3.12 is an EEPROM test program.

3.9   PIC16 C Analog Output

- Waveform generator test system

- Waveform test program

- Waveform output

In microcontroller applications, analog output is not needed as often as analog input, so no digital to analog converter (DAC) is built into the PIC MCU. An external DAC is needed to generate analog output signals.

A serial DAC may be used to output a precision DC reference voltage or low-frequency analog signal, using SPI or I^{2}C to transfer the data. A 10-bit or 12-bit output is typically provided, giving a precision of about 0.1 or 0.025%, respectively. However, the serial data transfer is inherently slow. In the demo system described here (Figure 3.12), higher speed is possible with parallel output to the DAC. The waveform generator circuit generates trigonometric waveforms, which are displayed on the virtual digital oscilloscope.

The system provides 8-bit conversion, giving a precision of $100/256 \approx 0.4\%$. With a 20-MHz MCU clock, the maximum output frequency is about 4 kHz. This is limited by the maximum rate at which the output loop can produce the instantaneous voltages that make up the waveform.

The DAC code is output at Port D, with a variable delay to control the frequency. A set of switches provides waveform selection and push-button frequency adjustment. The DAC0808 produces a current output that needs an external amplifier to convert it to a voltage and provide the output drive. The amplifier stage also allows the output amplitude and offset to be adjusted.

Listing 3.12   EEPROM Test Program

```c
// EEPROM.C
// Internal data EEPROM test, design file EEPROM.DSN

#include "16F877A.h"
#use delay(clock=4000000)

void main()  /////////////////////////////////////////////////////////
{
  int writebyte, readbyte;
  int maxadd, address;

  port_b_pullups(1);     // Enable Port B internal pull-ups
  if(!input(PIN_C1))     // Write memory sequence //////////////////
  {
    address=0;                                // First address

    do
    { while(input(PIN_C0)){};                 // Wait for button
      writebyte = input_B();                  // Get switch bank data
      write_eeprom(address,writebyte);        // Write data to EEPROM
      readbyte = read_eeprom(address);        // Read it back
      output_D(readbyte);                     // Display data on bar graph
      while(!input(PIN_C0)){};                // Wait for button release
      address++;                              // Next EEPROM address

    } while(writebyte!=0);                    // Continue until data = 00
  }

  else                   // Read memory sequence ///////////////////
  {
    address = 0;                              // First address

    do
    { while(input(PIN_C0)){};                 // Wait for button
      readbyte = read_eeprom(address);        // Read data
      output_D(readbyte);                     // Display it on bar graph
      while(!input(PIN_C0)){};                // Wait for button release
      address++;                              // Next address
    } while(readbyte!=0);                     // Continue until data = 00

  while(1);   // Done *************************************
  }
}
```

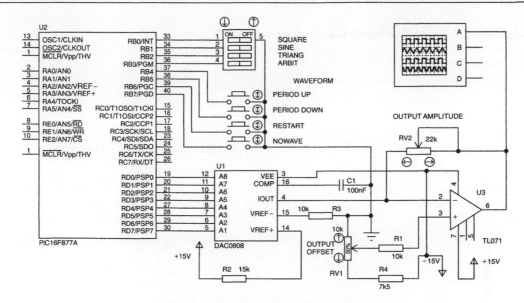

Figure 3.12: Waveform Generator

The program source code is shown in Listing 3.13. This is only a demonstration of the digital waveform generator principle, and a more sophisticated design is required to produce a waveform with a better resolution at higher frequencies. It serves only to illustrate some relevant features of C and the principle of waveform synthesis that may be used in high-performance digital signal processors, such as the dsPIC range. This is an application where critical sections of the code could be written in assembler for higher speed.

The main object of the program is to generate instantaneous voltages in sequence to produce a square, sine, triangular, and arbitrary waveform. The mid-value for the output is 100_{10}. Instant values ranging between -100 and $+100$ are added to this value to produce the output.

For the arbitrary pattern, most values are 0 in this example, with an increasing value at intervals of ten steps. This produces a pulse-modulated triangular waveform, which might be used to test a digital filter, but any other repetitive pattern can be entered as required. The arbitrary sequence is generated from the values entered into the array amp[n] in the function setwave() at the source code edit stage. A mechanism for entering these externally in hardware could easily be added, but that is rather tedious to demonstrate.

For the other waveforms, the values are calculated. The square wave is just a set of constant maximum ($+100$) and minimum (-100) values, and the triangular wave is an

Listing 3.13   Waveform Generator Source Code

```
// DACWAVE.C MPB 5-7-07
// Outputs waveforms to DAC, simulation file DAC.DSN

#include "16F877A.H"
#include "MATH.H"
#use delay(clock=20000000)
#use fast_io(D)                          // High speed output functions

int n, time=10;
float step, sinangle;
float stepangle = 0.0174533;         // 1 degree in radians
int amp[91];                         // Output instant voltage array

// ISR to read push buttons *******************************************

#int_rb
  void change()
  {
    if(time!=255)
      {if (!input(PIN_B4)) time ++;}            // Increase period
    while(!input(PIN_B4));

    if(time!=0)
      {if (!input(PIN_B5)) time--;}            // Decrease period
    while(!input(PIN_B5));

    if(!input(PIN_B6))reset_cpu();             // Restart program
    if(!input(PIN_B7))for(n=0;n<91;n++)amp[n]=0;  // Zero output
  }

void setwave()  // Arbitrary waveform values *********************
{
  amp[0]  =00;amp[1]  =00;amp[2] =00;amp[3] =00;amp[4] =00;
  amp[5]  =00;amp[6]  =00;amp[7] =00;amp[8] =00;amp[9] =00;
  amp[10] =10;amp[11] =00;amp[12]=00;amp[13]=00;amp[14]=00;
  amp[15] =00;amp[16] =00;amp[17]=00;amp[18]=00;amp[19]=00;
  amp[20] =20;amp[21] =00;amp[22]=00;amp[23]=00;amp[24]=00;
  amp[25] =00;amp[26] =00;amp[27]=00;amp[28]=00;amp[29]=00;
  amp[30] =30;amp[31] =00;amp[32]=00;amp[33]=00;amp[34]=00;
  amp[35] =00;amp[36] =00;amp[37]=00;amp[38]=00;amp[39]=00;
  amp[40] =40;amp[41] =00;amp[42]=00;amp[43]=00;amp[44]=00;
  amp[45] =00;amp[46] =00;amp[47]=00;amp[48]=00;amp[49]=00;
  amp[50] =50;amp[51] =00;amp[52]=00;amp[53]=00;amp[54]=00;
  amp[55] =00;amp[56] =00;amp[57]=00;amp[58]=00;amp[59]=00;
  amp[60] =60;amp[61] =00;amp[62]=00;amp[63]=00;amp[64]=00;
  amp[65] =00;amp[66] =00;amp[67]=00;amp[68]=00;amp[69]=00;
```

```
  amp[70]=70;amp[71]=00;amp[72]=00;amp[73]=00;amp[74]=00;
  amp[75]=00;amp[76]=00;amp[77]=00;amp[78]=00;amp[79]=00;
  amp[80]=80;amp[81]=00;amp[82]=00;amp[83]=00;amp[84]=00;
  amp[85]=00;amp[86]=00;amp[87]=00;amp[88]=00;amp[89]=00;
  amp[90]=90;
}

void main()  //*************************************************
{
  enable_interrupts(int_rb);        // Port B interrupt for buttons
  enable_interrupts(global);
  ext_int_edge(H_TO_L);
  port_b_pullups(1);
  set_tris_D(0);

  // Calculate waveform values ************************************

  step=0;
  for(n=0;n<91;n++)
  {
    if(!input(PIN_B0)) amp[n] = 100;      // Square wave offset
    if(!input(PIN_B1))                    // Calculate sine values
    { sinangle = sin(step*stepangle);
      amp[n] = floor(sinangle*100);
      step = step+1;
    }
    if(!input(PIN_B2)) amp[n] = n;        // Triangular wave
    if(!input(PIN_B3)) setwave();         // Arbitrary wave
  }

  // Output waveform vales ***************************************

  while(1)
  { for(n=0;n<91;n++)  {output_D(100+amp[n]); delay_us(time);}
    for(n=89;n>0;n--)  {output_D(100+amp[n]); delay_us(time);}
    for(n=0;n<91;n++)  {output_D(100-amp[n]); delay_us(time);}
    for(n=89;n>0;n--)  {output_D(100-amp[n]); delay_us(time);}
  }
}
```

incrementing and decrementing count. The sine output is the most interesting, as it is calculated using the sine function from the math.h library. These values are assigned to the amp[n] array for output after being calculated, since to calculate each and output it "on the fly" would be too slow.

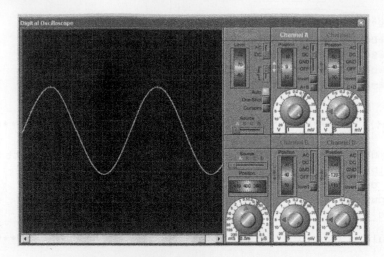

Figure 3.13: Sine Wave DAC Output

The waveform is selected at the start of the program by polling the selection switch bank. If the waveform selection is changed, the loop must be restarted using the push button. On the other hand, the frequency may be modified while the output is running. The main consideration here is the timing of the output waveform—each step must take the same time. The minimum step time is also important, as this determines the highest frequency. Therefore, input polling is avoided. Instead, the Port B change interrupt is used to detect the push buttons, and the period modification and waveform control operations are placed in the interrupt routine void change(). Here, the delay between each output step is incremented or decremented or the loop stopped and restarted. The sine waveform obtained is illustrated in Figure 3.13.

Assessment 3
5 points each, total 100

1. Write a C statement that sets up the PIC ADC so that only RA0 is used as an analog input. Deduce the resolution per bit for a 10-bit conversion, assuming a 5V supply.

2. If a single 4.096V reference voltage is connected to V_{ref+} and 10-bit conversion completed, write a C statement (a) to declare suitable variables and (b) to convert the input value to the actual voltage for display.

3. List the statements required to set up an ADC interrupt and outline the related ISR initialization if it is called "isrADC."

4. Explain the advantages of using an interrupt to read the data from an analog input conversion, compared with simply checking it on a regular basis (polling) within the program loop.

5. A 16-bit timer is preloaded with a value of 15,536. The MCU clock runs at 8 MHz, with a prescaler set to divide by 16. Calculate the timer output interval obtained.

6. Explain briefly the difference between the Capture and Compare modes of operation.

7. Draw a labeled diagram to show a PWM waveform, indicating how the overall period and duty cycle are set by the arguments of functions `setup_timer_2(a,b,c)` and `set_PWMx_duty(d)`. The MCU instruction clock period is T.

8. Calculate modified parameters for the setup functions in program PWM that produce an output at 1 kHz with a duty cycle of 10% (0.1-ms pulse). The instruction clock is 1 MHz.

9. Explain why the UART is a suitable interface for transmission of characters to a serial LCD display, especially if the LCD is separated from the MCU board.

10. Explain the effect of the statements `printf("%d",incode)` and `putc(incode)` on an LCD display connected to an MCU serial output, if the value of `incode` is `0x41`.

11. Outline how to structure a program using interrupts that can carry on some other task while the serial data are transferred to and from the UART, and explain why this might be useful.

12. By reference to Section 1.4, explain briefly how the hardware and master program would be modified if more than one slave sender were in the SPI system shown in Figure 3.8.

13. List the sequence of I^{2}C statements to write the data byte `0xAA` to address `0x01FF` in the serial memory chip in the system shown in Figure 3.9.

14. Draw a block diagram showing how to connect two PIC MCUs using an I^{2}C link.

the input is a maximum of 5 V, the value received by the ADC will be 256. The scaling factor therefore is 5/256 = 0.5 mV/bit. The input therefore needs to be multiplied by 0.0195 to be displayed as voltage. Floating point variables need to be used in the revised program

Assignment 3.2

Download the project PWM and test it for correct operation. A 250-Hz (4-ms) pulse waveform with a 50% duty cycle should be observed on the display. Now, rewrite the program to produce the same output using compare mode in 'Timer1'. The timer needs to run for 2ms for each half cycle; assuming a 4-MHz MCU clock and a 1-MHz timer clock, a compare value of 2000 is needed

Assignment 3.3

Download the project DACWAVE and test it for correct operation. Measure the minimum and maximum frequencies available. Modify the arbitrary-waveform data to produce a step waveform that has amplitude 0 for five steps of the output, 5 for the next five steps, 10 for the next five steps, and so on until the amplitude reaches 90 over the last five steps, then reduces to 0 again. It should then produce the same over the negative half cycle of the waveform before repeating.

PART 4

C Mechatronics Applications

4.1   PICDEM Mechatronics Board Overview

- Mechatronics board hardware
- Mechatronics board connections
- Mechatronics board motor drives

The PICDEM mechatronics demonstration board (Figure 4.1), supplied by Microchip®
Inc., is a very useful target system for C control applications. A user manual, which can be
downloaded from www.microchip.com, contains the schematics and general guidance on
using the board. It can be programmed using the ICD2 In-Circuit Debugger module, which
allows a final stage of fault finding when testing an application in the target hardware.
Alternatively, the low-cost PicKit2 programmer can be used. Since our applications here
have been tested in simulation mode, the full ICD debugging interface is not needed.

PICDEM Hardware

The block diagram, Figure 4.2, shows the main parts of the mechatronics board. It is built
around a PIC 16F917, which is similar to the 16F877A but incorporates an LCD driver
module, which allows a plain 3.5-digit display to be operated with no additional interfacing.

The MCU internal clock runs at 8 MHz, giving a 0.5-μs instruction cycle. The main
output devices are a small DC motor and a stepper motor. These are operated from a set
of four current driver FETs, which can sink or source current. These allow either motor
to be driven in both directions when connected as a full bridge driver. Input tactile push
switches and output LEDs are provided for simple test programs, mode selection, and

Figure 4.1: PICDEM Mechatronics Board (by permission of Microchip Inc.)

status indication. An RS232 serial port for exchanging data with the PC host is fitted, which requires a suitable terminal program running on the PC.

A temperature sensor is fitted, which outputs 10 mV/C with 0C giving 500 mV. Therefore, at 20°C, the output will be $500 + (20 \times 10) = 700$ mV. This voltage can be fed to an ADC input or comparator input on the MCU. A light sensor is also available, giving an output in the range 0–5 V. Two pots, giving 0–5 V, can be used as reference inputs for the analog sensors or as test inputs for analog applications.

The mechatronics board has its main signals brought to in-line connectors, as shown in the board layout (Figure 4.3). The components can be connected up for different applications using link wires. The connector pin functions are listed for reference in Tables 4.1 through 4.4.

Motor Drives

The motors are driven from a set of four half-bridge driver stages, which can handle up to 1 A each. These can be connected to the 5-V regulated or the 9–12-V unregulated supply for higher power output. Note that the main plug supply may be rated at less than 1 A, so a separate supply is advisable if the full drive current is needed.

Each driver has a pair of MOSFETs, which allow the stage to source or sink current, depending on which transistor is switched on (Figure 4.4). Control logic prevents both coming on at the same time and shuts down all the drives if an overcurrent fault is detected. This is activated on power up for fail safe operation and must be cleared manually before testing a motor. If the DC motor needs to be driven in both directions, the half-bridge

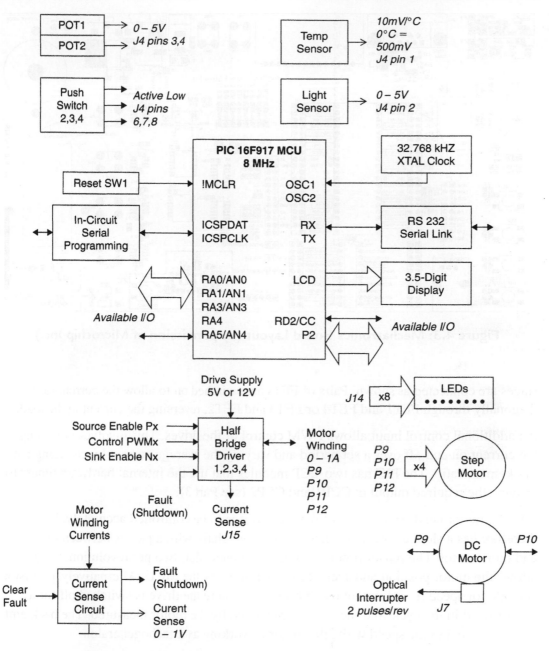

Figure 4.2: Block Diagram of PICDEM Mechatronics Board

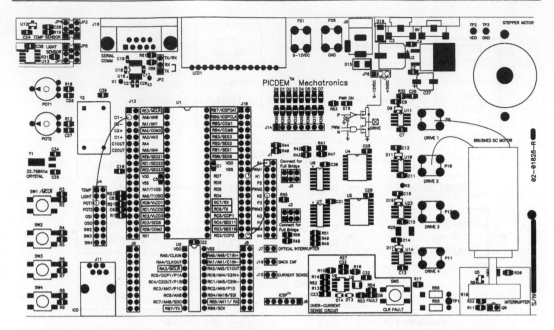

Figure 4.3: Mechatronics Board Layout (by permission of Microchip Inc.)

stages are connected as shown. Pairs of FETs are switched on to allow the current to flow diagonally through FET1 and FET4 or FET3 and FET2, reversing the current in the load.

An additional control input allows PWM control of the drives. This involves switching the current on and off over a set period and varying the average current by changing the mark-space ratio. The PIC has two CCP modules that use the internal hardware timers to provide the required output at CCP1 and CCP2 (see Part 3).

The DC motor needs some form of feedback if it is to be controlled accurately. It therefore has a slotted wheel attached to its output shaft, which passes between an LED and opto-sensor. The sensor produces a pulse for each slot, two per revolution, which allows the motor position and speed to be measured by the MCU. Alternatively, provision is made for speed measurement using back emf, where the drive is switched off for a short period in the cycle and the voltage generated by the motor measured. The back emf is proportional to the speed while the motor is working as a tachogenerator.

The stepper motor has two sets of windings, which are activated in sequence. This moves the rotor one step at a time, or 7.5 degrees. The windings are connected to separate full-bridge drivers consisting of half-bridges 1/2 and 3/4.

Table 4.1: Mechatronics Board Fixed Connections

Label	Alt Func	MCU Pin	Function
Dedicated I/O			
SW1 / !MCLR	RE3	1	Reset MCU (if enabled in fuses)
ICSPDATA	RB7	40	In-circuit serial programming data
ICSPCLK	RB6	39	In-circuit serial programming clock
RX	RC7	26	Receive data from RS232 interface
TX	RC6	25	Transmit data to RS232 interface
Display I/O			
SEG0	RB0	33	LCD segment 0 (see display map)
SEG1	RB1	34	LCD segment 1 (see display map)
SEG2	RB2	35	LCD segment 2 (see display map)
SEG3	RB3	36	LCD segment 3 (see display map)
SEG6	RC3	18	LCD segment 6 (see display map)
SEG21	RE0	8	LCD segment 21 (see display map)
SEG22	RE1	9	LCD segment 22 (see display map)
SEG23	RE2	10	LCD segment 23 (see display map)
COM0	RB4	37	LCD Common connection 0
COM1	RB5	38	LCD Common connection 1
COM2	RA2	4	LCD Common connection 2
COM3	RD0	19	LCD Common connection 3
VLCD1	RC0	15	LCD control voltage 1 ($V_{dd}/3 = 1.66$ V)
VLCD2	RC1	16	LCD control voltage 2 ($2\,V_{dd}/3 = 3.33$ V)
VLCD3	RC3	17	LCD control voltage 3 V_{dd}

Table 4.2: Mechatronics Board User Connections

User input devices	
SW2 SW3 SW4	General purpose tactile switches (active low), use RA0, RA1, RA3, RA4, RA5, RA6, RA7
POT1 POT2	Manual analog input (0–5V) for ADC and comparator, use AN1 (C1−), AN2, AN3 (C1+), AN4
Sensor inputs	
TEMP LIGHT	Temperature sensor (10mV/°C, 0°C = 500 mV), use AN1–AN4 Light sensor (0–5 V), use C1− and C1+

Table 4.3: DC Motor Connections

Label	Alt Func	MCU Pin
DC motor output (J1)		
P1	RD7	Enable source current driver stage 1
PWM1	CCP1	Pulse width control driver stage 1
N1	RD6	Enable sink current driver stage 1
P2	RD5	Enable source current driver stage 2
PWM2	CCP2	Pulse width control driver stage 2
N2	RD4	Enable sink current driver stage 2
DC motor sensors		
OPTINT	J7	Optical interrupter, 2 pulses per rev, use CCP1
BACKEMF	J16	Back EMF, 0–5 V, use RA1
CSENSE	J15	Current measurement, 1 mV/mA, use RA1

All bridge drives are connected to ground via a 0.1-Ω current sensing resistor, which produces a voltage proportional to the load current. This is fed to an amplifier and comparator so that the current can be measured. The comparator triggers a "fault" condition if the current exceeds 1 A (100 mV across the sensing resistor), which shuts down the drives. This fault condition also occurs on power-up, ensuring that the drives start only after the Clear Fault switch is pressed.

Table 4.4: Stepper Motor Connections

Label	Alt Func	MCU Pin
P1	RD7	Enable source current driver stage 1
PWM1	CCP1	Pulse width control driver stage 1
N1	RD6	Enable sink current driver stage 1
P2	RD7	Enable source current driver stage 2
PWM2	CCP1	Pulse width control driver stage 2
N2	RD6	Enable sink current driver stage 2
P3	RD5	Enable source current driver stage 3
PWM3	CCP2	Pulse width control driver stage 3
N3	RD5	Enable sink current driver stage 3
P4	RD5	Enable source current driver stage 4
PWM4	CCP2	Pulse width control driver stage 4
N4	RD4	Enable sink current driver stage 4

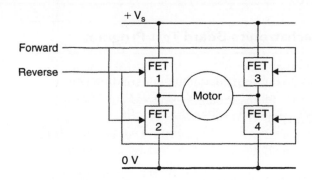

Figure 4.4: Full-Bridge Driver Connection of the DC Motor

Test Program

An initial test program for the PICDEM board is used to check that the downloading and in-circuit debugging modes are operational. The system setup is shown in Figure 4.5, the test program outline in Listing 4.1, and the source code in Listing 4.2. The program outline can be used in more complex applications to help to construct the program.

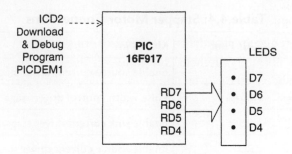

Figure 4.5: Block Diagram of Test Hardware Configuration

Listing 4.1   Test Program Outline

```
TEST
  Include 16F917 header file
  Use delay library routines

  Count = 0
  Loop always
    Output count at Port D
    Delay 10 ms
    Increment count
```

Listing 4.2   Mechatronics Board Test Program

```
//TEST.C MPB 14-4-07
//First program for testing Mechatronics Board
//Flashes 4 LEDs, total cycle time = 256 × 10 ms = 2.56 s
//Connect RD7-D7, RD6-D6, RD5-D5, RD4-D4

#include "16F917.h"              // Device header file
#use delay(clock=8000000)        // Delay function  clock speed

void main()                      //Start main block
{
  int n=0;                       //Count loop variable

  while(1)                       //Endless loop
  {
    output_D(n);                 //Show on LEDs
    delay_ms(10);                //Wait 10ms between steps
    n++;                         // Increment loop count
  }
}                                //End of source code
```

A program implements a simple output loop, which increments the binary count at Port C. The PIC 16F917 outputs RD4 to RD7 need to be connected to the LEDs D3 to D7 on the target board with link leads on the connector pins. The ICD2 module is plugged into the board via the ICD connector and to a host PC USB port.

The source code is loaded or edited in the usual way within MPLAB and saved in a project folder called "test." The source code and device header file are placed in the project folder and attached to the project in the project file window. Assuming the C compiler has been previously installed, the project can be complied and the HEX and COF files created.

The program is downloaded by selecting the menu Programmer, Select Programmer, MPLAB ICD2. Confirmation that the target is ready should appear in the output window. Hit the Program Target Device button and ideally a Programming Succeeded message is returned. The Release from Reset button should set the output running on the LEDs on the mechatronics board.

Debugging

If a program does not function correctly, it can be debugged in hardware using ICD2. For this exercise, we run the program in debug mode anyway. From the Debugger menu, Select Tool MPLAB ICD2. If necessary, the operating system in the ICD module is updated. A reminder may be received that the ICD2 module cannot operate as a debugger and programmer at the same time. An error message may be displayed at this stage, indicating that the system cannot enter debug mode. Resend the program and try again. The output window should then show that the target system is ready.

The debug control panel now appears in the toolbar, allowing the program to Run, Stop, Reset, or Single Step. The current execution point is displayed in the source code window. Reset the program if necessary, and run it. The LEDs should flash in a binary sequence on the target board. Stop the program and set a breakpoint at the output statement in the source code. Open the watch window and display the value of 'n' in binary. It increments each time the loop is executed, but note that the output shows only the most significant 4 bits. It therefore changes only after a count of 16.

You will find that the step-over function does not work. This is probably because the subroutine calls in CCS C are implemented using the assembler instruction GOTO instead of CALL, which the step-over function is expecting. This can be confirmed by opening the Disassembly Listing in the View menu.

Figure 4.6: Test Program Debugging Screen

The debug windows are shown in Figure 4.6. When debugging is complete, clear all breakpoints and ensure that the program is working as required. After the final version is downloaded and the ICD module disconnected, the program should run from Reset on power-up.

4.2   PICDEM Liquid Crystal Display

- LCD layout and connections
- LCD test program
- BCD count program

The plain 3.5-digit parallel liquid crystal display (LCD) is driven directly from the MCU, occupying 15 of the I/O pins. The usual alternative to this arrangement is to use a serial LCD, which can be driven via the RS232 port. This occupies only one or two pins, but it is more expensive, as it contains its own microcontroller.

LCD Connections

The parallel LCD is operated by specific combinations of inputs that enable the segments as required (Figure 4.7). The segments are designated A to G for each seven-segment

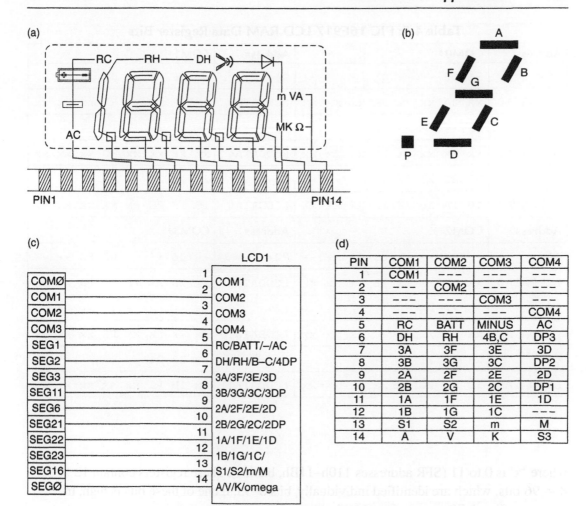

Figure 4.7: (a) LCD Segment Connections (courtesy of Varitronix Ltd.); (b) Segment Labels; (c) MCU to LCD Connection; (d) LCD Connection Map

digit, with digits numbered 1 to 4 from the right. The most significant half digit (4) has only segments B and C, displaying only '1'. Four common connections (COM1–COM4) enable groups of segments such that each has a unique address.

Note that this is a standard DMM display, so additional symbols are available that are not needed in the mechatronics board applications. The data for the display segments are stored in dedicated set of 12 registers in the PIC 16F917 (Table 4.5), called LCDDATAx,

Table 4.5: PIC 16F917 LCD RAM Data Register Bits

Address	COM0+								Address	COM1+							
00+	--	06	--	--	03	--	--	--	24+	--	06	--	--	03	--	--	--
LCDDATA0	xx	2A	xx	xx	3A	xx	xx	xx	LCDDATA3	xx	2F	xx	xx	3F	xx	xx	xx
08+	--	--	--	--	11	--	--	--	32+	--	--	--	--	11	--	--	--
LCDDATA1	xx	xx	xx	xx	3B	xx	xx	xx	LCDDATA4	xx	xx	xx	xx	3G	xx	xx	xx
016+	23	22	21	--	--	--	--	--	40+	23	22	21	--	--	--	--	--
LCDDATA2	1B	1A	2B	XX	XX	XX	XX	XX	LCDDATA5	1G	1F	2G	xx	xx	xx	xx	xx
Address	**COM2+**								**Address**	**COM3+**							
48+	--	06	--	--	03	02	--	--	72+	--	06	--	--	03	02	--	--
LCDDATA6	xx	2E	xx	xx	3E	4x	xx	xx	LCDDATA9	xx	2D	xx	xx	3D	P3	xx	xx
56+	--	--	--	--	11	--	--	--	80+	--	--	--	--	11	--	--	--
LCDDATA7	xx	xx	xx	xx	3C	xx	xx	xx	LCDDATA10	xx	xx	xx	xx	P2	xx	xx	xx
64+	23	22	21	--	--	--	--	--	88+	--	22	21	--	--	--	--	--
LCDDATA8	1C	1E	2C	xx	xx	xx	xx	xx	LCDDATA11	xx	1D	P1	xx	xx	xx	xx	xx

where 'x' is 0 to 11 (SFR addresses 110h–11Bh, bank 1). These registers contain $12 \times 8 = 96$ bits, which are identified individually, bits 0–95. If one of these bits is high, the corresponding LCD segment or pixel is on.

The LCD has a total of 26 numerical segments, comprising three seven-segment digits, two segments for the MSD, and three decimal points. The MSD bits are controlled by the same bit, as they always come on together, giving only 25 bits actually required. Therefore, only some bits in the registers are used, but the spare capacity allows more complex displays to be operated by the '917 in other applications. We see that the bits that are used are not arranged very logically, so they will be mapped by the LCD display function to simplify the output process.

The bits in the first three registers (LCDDATA0–LCDDATA2) are associated with COM0 output, the next three with COM1, and so on to COM3 (see Table 4.5). Unfortunately, the common inputs on the LCD are identified as COM1–COM4, so COM1 is controlled from the MCU output COM0, and so on, with COM4 being connected to COM3 MCU output pin.

The 16F917 MCU can provide up to 24 segment drive outputs (SEG0–SEG23), with four common connections (COM0–COM3). These are used in defined combinations to control up to 24 × 4 = 96 segments or pixels in the display. In this way, 1 bit in the LCDDATAx registers controls one element of the display. This display needs only 25 bits and ten of the available segment outputs (SEG0, 1, 2, 3, 6, 11, 16, 21, 22, and 23). These outputs are encoded to allow individual bit control within the program.

LCD Test Program

Listing 4.3 shows a test program, LCD1, which displays the numerals 0 to 9 on each digit in turn, then flashes on the MSD and three decimal points, so that correct operation of each can be checked.

Listing 4.3 Test Program for Mechatronics Board LCD

```
// LCD1.C MPB 20-4-07
// Test program for mechatronics board LCD
// Displays count 0 to 9 on Digits1,2,3 and 1 on Digit4

#include "16F917.h"
#use delay(clock=8000000)

//LCD DISPLAY DATA: (3 numerals*7 segments)+MSD*1 segment+3 decimal
points
//Bit map for numerals 0-9 and blank...............................
//Numeral:          0   1   2   3   4   5   6   7   8   9   blank
byte const DigMap[11]={0xFD,0x60,0xDB,0xF3,0x66,0xB7,0xBF,0xE0,0xFF,0xE7,0x00};

//Bit addressess in LCD RAM locations LCDDATA0 to LCDDATA11=12*8 bits
//Numbered 0-95 with offsets COM0=0, COM1=24, COM2=48, COM3=72
//Segment:      A        B        C        D        E        F        G
#define DIG1 COM0+22,COM0+23,COM2+23,COM3+22,COM2+22,COM1+22,COM1+23
//Bit addresses
#define DIG2 COM0+6, COM0+21,COM2+21,COM3+6, COM2+6, COM1+6, COM1+21
//Bit addresses
#define DIG3 COM0+3, COM0+11,COM2+11,COM3+3, COM2+3, COM1+3, COM1+11
//Bit addresses
```

```c
#define DIG4 COM2+2             //Both bits
#define DP1 COM3+21            //Decimal point 1
#define DP2 COM3+11            //Decimal point 2
#define DP3 COM3+2             //Decimal point 3
void main()
{
  int8 n;
  setup_lcd(LCD_MUX14,0);      //Initialize 14-pin LCD, no clock
                                 divide

  for(n=0;n<11;n++)            //Display numerals 0-9 at digit 1
  { lcd_symbol(DigMap[n],DIG1); //Send digit bits to segment
                                 addresses

    delay_ms(300);
  }

  for(n=0;n<11;n++)            //Display numerals 0-9 at digit 2
  { lcd_symbol(DigMap[n],DIG2); //Send digit bits to segment
                                 addresses

    delay_ms(300);
  }

  for(n=0;n<11;n++)            //Display numerals 0-9 at digit 3
  { lcd_symbol(DigMap[n],DIG3); //Send digit bits to segment
                                 addresses

    delay_ms(300);
  }

  lcd_symbol(0X80,DIG4);       //Switch on MSD digit 4
  delay_ms(1000);
  lcd_symbol(0X00,DIG4);       //Switch off MSD digit 4
  lcd_symbol(0XFF,DP1);        //Switch on decimal point 1
  delay_ms(500);
  lcd_symbol(0X00,DP1);        //Switch off decimal point 1
  lcd_symbol(0XFF,DP2);        //Switch on decimal point 2
  delay_ms(500);
  lcd_symbol(0X00,DP2);        //Switch off decimal point 2
  lcd_symbol(0XFF,DP3);        //Switch on decimal point 3
  delay_ms(500);
  lcd_symbol(0X00,DP3);        //Switch off decimal point 3

  while(1){};                  //Done

}
```

Table 4.6: Bit Maps for LCD Numerals

Numeral	Segment	A	B	C	D	E	F	G	LSB	Code
	Bit	7	6	5	4	3	2	1	0	
0		1	1	1	1	1	1	0	1	0xFD
1		0	1	1	0	0	0	0	0	0x60
2		1	1	0	1	1	0	1	0	0xDB
3		1	1	1	1	0	0	1	1	0xF3
4		0	1	1	0	0	1	1	0	0x66
5		1	0	1	0	0	1	1	1	0xB7
6		1	0	1	0	1	1	1	1	0xBF
7		1	1	1	0	0	0	0	0	0xE0
8		1	1	1	1	1	1	1	1	0xFF
9		1	1	1	0	0	1	1	1	0xE7
–		0	0	0	0	0	0	0	0	Blank

Each group of segments associated with each common connection on the LCD is operated in turn by the program. The LCD functions used are `setup_lcd()` and `lcd_symbol()`. The arguments of the setup function specify a 14-pin display module and 0 clock divide factor. The clock rate controls the display multiplexing rate, which can be modified for best visibility.

The arguments of the output function comprise an 8-bit map for the numeral to be displayed as a hex number (Table 4.6) and a list of the corresponding bits in the `LCDDATAx` locations for that digit. The 8-bit numeral codes are shown in Figure 4.7. Because of the interaction of the control lines, the LSB for each code was determined by inspecting the results on the display. Otherwise, the mapping is as normally required for seven segment codes.

The mapping data for each segment is provided to the output function in the form of a list of segment bit addresses, 0–95. To include information about which COM line is active for each bit, the address is supplied as the sum of the start address of each COM block and the bit number within that block. Therefore, the bit address of segment A of digit 1 (`DIG1`) is COM0 + 22. COM0 has the value 0, COM1 = 24, COM2 = 48, and COM3 = 72. Therefore, COM0 + 22 = 22. By the same process, the single-bit address controlling the MSD (DIG4) is COM2 + 2 = 50, and the first decimal point (DP1) is addressed at COM3 + 21 = 93.

For convenience, the lists of segment bit addresses for each digit are defined at the top of the program, using the replacement text labels DIG1, DIG2, DIG3, and DIG4 plus the three decimal point addresses. The lcd_symbol() function is then supplied with the constant array element number for the numeral to be displayed and the bit address list as DIGx. A 'for' loop outputs each numeral at each position in turn, including the blank digit, while the MSD and decimal points are switched on and off individually.

BCD Count Program

Listing 4.4 shows a program that displays a decimal count on the LCD. The count is generated as binary coded decimal (BCDx) digits. Each digit is initialized to 0, then incremented until it reaches 10, when it is cleared back to 0 and the next most significant digit incremented. The three digits are then displayed together. The MSD (DIG4) is not used. The LCD data block is now concealed in a separate source code file lcd.inc, which is included at the top of the program.

4.3   PICDEM DC Motor Test Programs

- Motor test program
- Rev counter program

The primary target device on the board is the DC motor. The hardware configuration is shown in Figure 4.8. The first program just switches the motor on and off, and the second shows how to control the speed.

Basic Control

The minimal program (Listing 4.5) shows how to run the mechatronics board under the control of SW2. The motor is connected to Drive1 and Drive2 output terminals, with two output bits of the MCU linked to P1 and N2. When these go high, the motor current is switched on in a forward direction. The output code 0x90 = 10010000_2 switches on RD4 and RD7 when the switch input RA4 goes low. If desired, the PIC output pins can also be monitored on the LEDs. The project should be loaded and tested as described in Section 4.1.

Listing 4.4 LCD Counting Program

```
/////////////////////////////////////////////////////////////////////
//LCD2.C MPB 20-4-07
//LCD program to count up when SW2 on
//Hardware: Connect SW2 to RA4
/////////////////////////////////////////////////////////////////////

#include "16F917.h"
#include "lcd.inc"                           //Include file with LCD data
#use delay(clock=8000000)

void main() /////////////////////////////////////////////////////////
{
  int8 BCD1=0, BCD2=0, BCD3=0;              //BCD count digits
  setup_lcd(LCD_MUX14,0);                   //Initialize 14-pin LCD

  while(1)
  {                                         //GENERATE DECIMAL COUNT
    if(!input(PIN_A4))                      //Test Switch 2
      {
        delay_ms(10);                       //Debounce and slow
        BCD1++;                             //Increment ones
        if(BCD1==10)                        //..up to 9
        {
          BCD1=0;                           //Reset ones
          BCD2++;                           //Increment tens
          if(BCD2==10)                      //..up to 90
          {
            BCD2=0;                         //Reset tens
            BCD3++;                         //Increment hundreds
            if(BCD3==10)                    //..up to 900
              BCD3=0;                       //All reset to zero
          }
        }
      }

                       //DISPLAY BCD DIGITS
    lcd_symbol(DigMap[BCD1],DIG1);          //Display Digit 1
    lcd_symbol(DigMap[BCD2],DIG2);          //Display Digit 2
    lcd_symbol(DigMap[BCD3],DIG3);          //Display Digit 3

  }                                         //Loop always

}/////////////////////////////////////////////////////////////////END
```

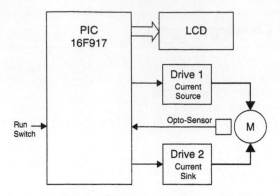

Figure 4.8: Block Diagram of Motor Test System

Listing 4.5   Motor Test Program

```
//MOTOR1.C MPB 17-4-07      PICDEM board test program
//Control motor from switch. Connect SW2-RA4, RD7-P1, RD4-N2

#include"16F917.h"

void main()
{
  while(1)
    {
      if(!input(PIN_A4))        //Test switch
        output_D(0x90);         //Switch on motor
      else output_D(0x00);      //Switch off motor
    }
}
```

Rev Counter

The system is now developed to measure the number of revolutions completed during a short run. The motor is still attached to Drive1 and Drive2 outputs, but in addition, the output from the opto-sensor (OPTO), which produces two pulses per rev, is connected to the Timer1 input on the MCU (RC5/T1CLKI). The motor is switched on by pressing SW2, and the number of revs is displayed when it is released. The maximum rev count is 999 (1998 pulses), which takes about 20 sec to reach, assuming the motor is running at about 3000 rpm. The program source code is given in Listing 4.6 and is outlined in Listing 4.7.

Listing 4.6 Program to Display Motor Revs

```
//////////////////////////////////////////////////////////////////////
//  MOTREVS.C
//  Program to count motor revs
//  PICDEM hardware: Connect SW2-RA4, RD4-N2, RD7-P1
//////////////////////////////////////////////////////////////////////

#include "16F917.h"
#include "lcd.inc"                        //Include file with LCD data
#use delay(clock=8000000)

void main() //////////////////////////////////////////////////////////
{
  int8 BCD1=0, BCD2=0, BCD3=0;            //Initialize 3 digits
  int8 huns=0, tens=0, ones=0;           //and digit values
  int16 count=0;                          //Receives timer count

  setup_lcd(LCD_MUX14,0);                 //Initialize 14-pin LCD
  setup_timer_1(T1_EXTERNAL);             //Initialize rev counter

  while(1)                                //Main loop start
  {
    while(input(PIN_A4)){};               //Wait for switch 2 on
    delay_ms(10);                         //Debounce switch

    //COUNT MOTOR REVSX2///////////////////////////////////////////////

    set_timer1(0);                        //Reset counter
    output_D(0x90);                       //Start motor
    while(!input(PIN_A4))                 //Wait while switch on
    {   delay_ms(10);  }                  //Debounce switch
    output_D(0x00);                       //Motor off
    count=get_timer1();                   //Read counter
    count=count/2;                        //2 pulses per rev

    //CONVERT COUNT TO BCD/////////////////////////////////////////////

    huns=tens=ones=0;                     //Reset digit values
    while (count>99)                      //Calculate hundreds
    {   count=count-100; huns++;  }       //digit by subtraction
    while (count>9)                       //Calculate tens
    {   count=count-10; tens++;   }       //digit by subtraction
    ones=count;

    //DISPLAY BCD DIGITS///////////////////////////////////////////////

    lcd_symbol(DigMap[ones],DIG1);        //Display Digit 1
    lcd_symbol(DigMap[tens],DIG2);        //Display Digit 2
    lcd_symbol(DigMap[huns],DIG3);        //Display Digit 3
  }                                       //Loop always
}    //////////////////////////////////////////////////////////////END
```

Listing 4.7   Outline of Rev Counter Program

```
MOTREVS
  Specify MCU 16F917
  Include LCD function file

  Initialize display digits to zero
  Setup LCD
  Setup timer as external pulse counter

  Main loop
    Display 3 digits on LCD
    Wait for input switch on
    Reset counter and start motor
    Wait for input switch off
    Stop motor
    Convert timer count to 3 digit BCD
```

The timer is set up for external input using `setup_timer_1(T1_EXTERNAL)`, and the resulting count is read using `get_timer1()`. The binary number obtained from the timer is divided by 2 and converted to BCD by a process of successive subtraction, which is simple if not elegant. The calculated digits are then displayed as in previous examples, using the function `lcd_symbol()` to output the display digits and the include file LCD.INC for the display encoding.

4.4   PICDEM Stepper Motor Control

- Stepper motor operation

- Stepper motor test program

- Speed and direction control

The main advantage of the stepper motor is that it provides position control without the feedback required by a DC motor. It has stator windings distributed around a cylindrical rotor, which has permanent or induced magnetic poles. The windings operate in groups to move the rotor by a fraction of a revolution at a time (Figure 4.8).

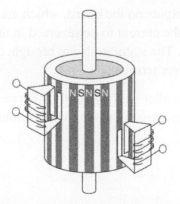

Figure 4.9: Bipolar Permanent Magnet Stepper Motor with Two Winding sets

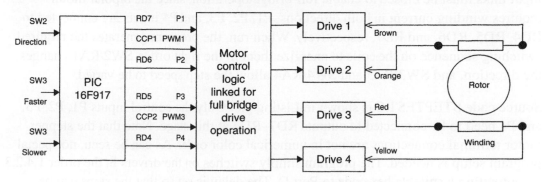

Figure 4.10: Stepper Motor Test System Connections

Construction

The small stepper motor on the mechatronics board is an inexpensive permanent magnet (PM) type, giving 7.5 degrees per step, 48 steps per revolution. It can also be moved in half steps by suitable operation of the windings or even smaller steps (microstepping) by suitable modulation of the winding current. The motor has two bipolar windings, which means the current is reversed to change the polarity of the stator pole. The coils energize two rings of poles, creating alternating north and south poles, which interact with the permanent rotor poles (Figure 4.9).

Representative windings are shown Figure 4.10; in the actual motor, coils are distributed around the whole circumference, multiplying the torque produced. Their terminals are connected to the four driver outputs on the board, which are normally connected for full-bridge operation. This allows the current to be reversed in the stator windings, reversing the polarity of the stator poles. The stator coils are brought out to four color-coded wires, which are connected to the driver terminals.

In more expensive motors, a smaller step (typically 1.8°) can be obtained with four sets of windings. These motors usually have six wires, with a common connection for each pair of windings.

Stepper Motor Test

The stepper motor is connected to the driver outputs, in clockwise order. The six driver input links must be closed to enable full-bridge operation, since the bipolar motor requires winding current in both directions. P1, P2, P3, and P4 inputs are connected to RD4, RD5, RD6, and RD7, respectively. When run, the program generates the required switching sequence on the coils to energize them in the right order. SW2/RA1 changes the direction, and SW3/RA3 and SW4/RA4 allow the step speed to be varied.

Source code STEPTEST.C is shown in Listing 4.8. Only the control inputs P1, P2, P3, and P4 need to be connected to outputs RD7–RD4 at this stage. Note that the stepper motor terminal connections are not in numerical color order. As can be seen, no special program setup is needed. The program simply switches on the drivers in the order 1,4,2,3 by outputting a suitable hex code to Port D. The delay is set so that the steps can be counted visually. It is helpful to attach an indicator flag to the motor shaft, so that the stepping can be seen more easily. The number of full steps per rev can then be confirmed (48).

Program STEPSPEED, Listings 4.9 and 4.10, is a development of the basic program to test the motor response to a range of step rates. The input tactile switches change the speed by modifying the delay time parameter, which is set to 16 ms by default. This gives speed of

$$16 \text{ ms/step} = 16 \times 48 = 0.768 \text{ sec/rev} = 0.768 \times 60 = 46 \text{ rpm}$$

Direction Control

The stepper motor program can now be further developed to include direction control, as shown in STEPDIR.C (Listing 4.11). The program has been restructured to incorporate

Listing 4.8 Stepper Motor Test Program

```
//  STEPTEST.C
//  Test program for PICDEM Mechatronics Board stepper motor,
//  basic full step mode. Connect RD7-P1, RD6-P2, RD5-P3, RD4-P4
//  plus all 6 jumpers for full bridge mode
//  Motor moves 48 steps per rev (7.5 deg/step)
/////////////////////////////////////////////////////

#include "16F917.h"
#use delay(clock=8000000)

void main()
{
  while(1)                 //Loop always
  {
    output_D(0x80);        //Switch on Drive 1
    delay_ms(200);

    output_D(0x10);        //Switch on Drive 4
    delay_ms(200);

    output_D(0x40);        //Switch on Drive 2
    delay_ms(200);

    output_D(0x20);        //Switch on Drive 3
    delay_ms(200);
  }
}
```

Listing 4.9 Outline of Stepper Motor Speed Control Program

```
STEPSPEED
  Specify MCU 16F917
  Set default step delay time

  Main loop
    If Direction switch pulsed, Call Forward
    If Direction switch pulsed, Call Reverse

    Forward
      Call Speed
      Output one forward cycle (4 steps) to motor

    Reverse
      Call Speed
      Output one reverse cycle (4 steps) to motor

      Speed
        If Up button on, halve step delay
        If Down button on, double step delay
```

Listing 4.10   Stepper Motor Speed Control Program

```
////////////////////////////////////////////////////////////////////
//  STEPSPEED.CMPB 22-4-07
//  Program for PICDEM Mechatronics Board stepper motor, full step mode
//  Connect RD7-P1, RD6-P2, RD5-P3, RD4-P4 plus all 6 jumpers for full
//  bridge mode plus SW3-RA3 and SW4-RA4. Motor speed SW3 up SW4 down
////////////////////////////////////////////////////////////////////

#include "16F917.h"
#use delay(clock=8000000)

void main()
{
  int8 time=16;                       // Variable step delay

  while(1)                            //Loop always
  {

                                      //CHECK SWITCHES

    if(!input(PIN_A3))                //Poll SW3
    {    delay_ms(10);                //Debounce
         if(time!=1)time=time/2;      //Not if min
    }
    while(!input(PIN_A3)){};          //Wait switch

    if(!input(PIN_A4))                //Poll SW3
    {    delay_ms(10);                //Debounce
         if(time!=128)time=time*2;    //Not if max
    }
    while(!input(PIN_A4)){};          //Wait switch

                                      //4 STEPS CLOCKWISE

    output_D(0x20); delay_ms(time);   //Step 1
    output_D(0x40); delay_ms(time);   //Step 2
    output_D(0x10); delay_ms(time);   //Step 3
    output_D(0x80); delay_ms(time);   //Step 4

  }
}
```

---

a procedure for modifying speed. In the main loop, the reversing button is tested; by default the motor runs forward and is reversed each time the button is pressed. Before each sequence of four steps, the speed buttons are polled and the delay modified if requested. The structure makes it easier to write the program with the right logical sequence. A flaw

---

Listing 4.11 Stepper Motor Speed and Direction Control

```c
/////////////////////////////////////////////////////////////////
//   STEPDIR.C PICDEM Mechatronics Board stepper motor speed and dirc.
//   Connect RD7-P1, RD6-P2, RD5-P3, RD4-P4 plus all 6 jumpers(full bridge)
//   SW2-RA2, SW3-RA3, SW4-RA4. Motor speed SW3 up SW4 down, motor dirc SW2
/////////////////////////////////////////////////////////////////

#include "16F917.h"                        //MCU select
#use delay(clock=8000000)                  //Internal clock
int8 time=16;                              //Default speed

//PROCEDURES//////////////////////////////////////////////////////

  void speed() //Halve or double speed //////////
    {
      if(!input(PIN_A3))                   //Poll SW3
      { delay_ms(10);                      //Debounce
        if(time!=1)time=time/2;            //Not if min
      }
      while(!input(PIN_A3)){};             //Wait switch

      if(!input(PIN_A4))                   //Poll SW3
      { delay_ms(10);                      //Debounce
        if(time!=128)time=time*2;          //Not if max
      }
      while(!input(PIN_A4)){};             //Wait switch
    }

  void forward() //4 steps clockwise /////////////
    {
      speed();
      output_D(0x20); delay_ms(time);      //Step 1
      output_D(0x40); delay_ms(time);      //Step 2
      output_D(0x10); delay_ms(time);      //Step 3
      output_D(0x80); delay_ms(time);      //Step 4
    }

  void reverse() //4 steps counter-clockwise /////
    {
      speed();
      output_D(0x80); delay_ms(time);      //Step 4
      output_D(0x10); delay_ms(time);      //Step 3
      output_D(0x40); delay_ms(time);      //Step 2
      output_D(0x20); delay_ms(time);      //Step 1
    }
```

```
void main() //Main loop/////////////////////////////////////////////
{
  while(1)                                   //Loop always
  {
    while(input(PIN_A2)) { forward(); }      //Run forward
    delay_ms(10);                            //Debounce
    while(!input(PIN_A2)){};                 //Wait until released

    while(input(PIN_A2)) { reverse(); }      //Run reverse
    delay_ms(10);                            //Debounce
    while(!input(PIN_A2)){};                 //Wait until released
  }
}
```

in the algorithm is that the program checks the buttons only after four steps, so the direction and speed do not change immediately if the motor is running at low speed. This type of problem can be solved using interrupts.

4.5   PICDEM Analog Sensors

- Light switch application

- Temperature display application

The mechatronics board is fitted with a light and temperature sensor, each of which produces an analog output in the range of 0–5 V. In common with many sensors now available, a signal conditioning amplifier is built in, so that no additional components are needed to interface with an MCU.

Light Sensor

The light sensor can be tested using the analog comparator inputs of the 16F917, which allow two input voltages to be compared. An output bit in a status register is set if the positive input (C+) is at a higher voltage than the negative input (C−) or a reference voltage. A range of setup options are defined in the header file.

The block diagram in Figure 4.11 shows the hardware configuration for this test. The connector pin LIGHT, the light sensor output, is connected to RA0 (comparator input C−) and POT1 to RA3 (comparator input C+), with LED D7 is assigned to RD7 to display the

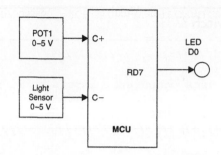

Figure 4.11: Comparator Test Setup

Listing 4.12   Outline of Light Sensor Test Program

```
LIGHTCON
  Select MCU 16F917
  Initialize comparator input

Main loop
  If light > set level, switch output OFF
  Else switch output ON
```

---

comparator state. When the light level is reduced, the output switches on. Conversely, it goes off as the light is increased through the switching level, which is adjustable using POT1. This simulates the operation of an automatic streetlight switch or security lamp. The program LIGHTCON is outlined in Listing 4.12 and the source code shown in Listing 4.13.

As we see, in the program, only the setup function is needed, which assigns the comparator inputs to Port A pins. Two comparators are available, and the setup used here is the same for all comparator applications using this hardware. C1OUT is the bit label assigned to the Comparator 1 output bit, which is tested using the `if` statement. The LED output is then switched accordingly. The pot sets the switching level, and a desk lamp or flashlight was found to work as a light source. The LED should go on when the light source goes off.

Temperature Measurement

The temperature sensor on the PICDEM board has an output of 10 mV/°C, with 500 mV = 0°C (Figure 4.12). For this application, the TEMP pin, to which the temperature sensor output is connected, is linked to the first analog input RA0 (AN0). When run, the

---

Listing 4.13   Light Switch

```
///////////////////////////////////////////////////////////////////
//  LIGHTCON.C
//  Auto light switch uses comparator inputs on mechatronics board
//  Pot 1 adjusted for light switching level.
//  Connect: LIGHT to C1-, POT1 to C1+
///////////////////////////////////////////////////////////////////

#include "16F917.h"

void main()
{
  setup_comparator(A0_A3_A1_A2);      //Setup for PICDEM board

  while(1)
  {  if(!C1OUT) output_low(PIN_D7);   //Switch off LED if light>pot
     else output_high(PIN_D7);        //Switch on LED if light<pot
  }
}
```

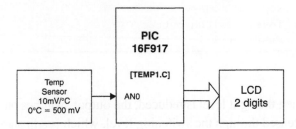

Figure 4.12: Temperature Sensor System

temperature is converted and displayed. The program TEMPDIS outline is given in Listing 4.14 and the source code in Listing 4.15.

The ADC is set to 10-bit conversion, giving an output of 1024 steps:

Internal ADC reference voltage = 5.00 V.
Bit resolution = 5.00/1024 = 4.88 mV per bit.
Temperature measurement = 10 mV per °C.
Temperature resolution = 4.88/10 = 0.488°C per bit.

The temperature is therefore measured to about 0.5°C. This is quite acceptable, as the display is precise to only ±1°C. By contrast, if 8-bit conversion were used, the precision would be only about 2°C per bit and the display would be misleading.

Listing 4.14 Outline of Temperature Sensor Test Program

```
TEMPDIS
  Select MCU 16F917
  Include LCD functions

  Setup LCD
  Setup ADC (10 bits, AN0)

  Main loop
    Read analogue input (binary 0-1024)
    Convert to temperature value (integer)
    Convert to BCD digits
    Display on LCD (0-99)
```

Listing 4.15 Temperature Display Source Code

```
//////////////////////////////////////////////////////////////////
//   TEMP1.C MPB 24-4-07
//   Demo program for PICDEM Mechatronics Board
//   Displays temperature +1/-0 deg C. Target board link: TEMP-AN0
//////////////////////////////////////////////////////////////////

#include "16F917.h"                    //MCU header file
#device ADC=10                         //Select 10-bit ADC
#include "lcd.inc"                      //LCD segment map file

void main()                            //Start main block
{
  int16 intemp;                        //Input temp from ADC result
  float temp;                          //Decimal result of scaling
  int8 distemp, tens, ones;            //Display temp and BCD digits

  setup_lcd(LCD_MUX14,0);              //Initialize 14-pin LCD
  setup_adc(ADC_CLOCK_INTERNAL);       //Select internal ADC clock
  setup_adc_ports(sAN0);               //Configure for AN0 input
  set_adc_channel(0);                  //Select AN0

  while(1)                             //Main loop always
  {
    intemp=read_adc();                 //Read analog input
    temp=(intemp*0.488)-50;            //Convert to degC
    distemp=temp;                      //Truncate to integer

    tens=temp/10;
    ones=distemp-(10*tens);            //Calculate BCD ones digit

    lcd_symbol(DigMap[ones],DIG1);     //Display low digit
    lcd_symbol(DigMap[tens],DIG2);     //Display high digit
  }
}
```

The program needs to convert the input to degrees C by multiplying the input bit count by the temperature resolution, 0.488°C per bit. Since the temperature range effectively starts at 0°C = 500 mV, we must subtract this offset from the calculated temperature. For example, at room temperature of 20°C, the sensor output is 500 + (20 × 10) = 700 mV. This converts to a value of 700/4.88 = 143 (nearest integer).

We check that we see the correct display:

$$(143 \times 0.488) - 50 = 19.8°C.$$

Due to rounding down in the program, this displays as 19°C and changes to 20°C only when this input has been exceeded, so the display shows the correct temperature accurate to +1°C and −0°C. A correcting factor of approximately +1/2°C could be implemented by simply adding 1 to the ADC result to give a display to the nearest whole degree.

Note that the automatic type conversion incorporated into the complier simplifies the arithmetic significantly. The type is changed automatically while preserving the value as far as is possible in the new format. Therefore, a decimal is truncated to an integer by simple assignment of the value from a float to integer variable.

4.6   PICDEM Temperature Controller

- Specification of temperature controller
- Input and output allocation
- Program outline

The PICDEM mechatronics board will now be used as the hardware platform for a temperature controller. Using a ready-made board eliminates the need for detailed hardware design and should be considered if a suitable product is available at a reasonable cost.

Specification

A temperature controller is required to control a greenhouse or similar outdoor enclosure at a temperature of 25–30°C using electric heaters and a cooling fan.

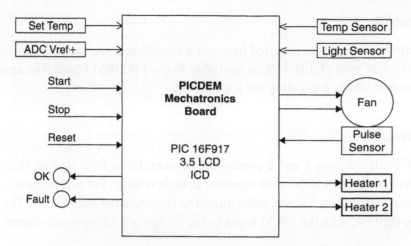

Figure 4.13: Block Diagram of the Temperature Controller

1. Overall function

   Maintain target temperature within $+/-2°C$, displaying it on the LCD. If the temperature is within specifications, switch on RunOK indicator; if temperature difference exceeds 5°C, switch on flash fault indicator.

2. Startup procedure

   - Power up the system, reset the fault indicator.

   - Display the set temperature on the LCD for operator adjustment.

   - Wait for the start input push button.

3. Overall operation

   - Switch on the first heater if the temperature is more than 2°C below the target.

   - Switch on the other heater if the temperature is more than 5°C below the target.

   - Run fans at a speed proportional to the positive temperature difference: If the fan speed is zero, switch on the fault indicator; if the temperature sensor is out of range, enable the fault mode.

   - If the light level indicates direct sunlight, add a positive offset to the fan speed in anticipation of an additional temperature rise. If the light sensor is out of range, enable the fault mode.

The block diagram, Figure 4.13, shows the system I/O requirements.

I/O Allocation

Once the inputs and outputs required have been established, we can provisionally assign them to particular pins (Table 4.7), as available in the PICDEM board. The appropriate links can later be made for testing the application.

Implementation

Output half-bridge drivers 1 and 2 control the heaters. In the final system, these are interfaced via contactors if the load operates at high voltage. For test purposes, a 6-V filament lamp is connected to the drive output to represent the heater load. The motor is operated by drive 4, with the PWM input to the bridge providing speed control. All these

Table 4.7: PICDEM Board I/O Allocation for Temperature Controller (Excluding LCD)

Pin	Label	Type	Board	Description
RA0	Tempin	Analog in	TEMP	Range 0–50°C = 500–1000 mV
RA1	Lightin	Analog in	LIGHT	Range 0–5 V, needs calibration
RA2	SetTemp	Analog in	POT1	Range 0.5–1.00V, set target temp
RA3	V_{ref+}	Analog in	POT2	Adjusted to 1.024V
RA5	Startin	Digital in	SW2	Active low, push button, start system
RA6	Stopin	Digital in	SW3	Active low, push button, shut down
RE3	Reset	Digital in	SW1	Active low (hard wired) !MCLR
RD4	RunOK	Digital out	D0	Active high, status indicator LED
RD5	Fault	Digital out	D1	Active high, status indicator LED
RD6	FanPWM	Digital out	PWM4	Active high, DC motor, DRIVE 4
RD7	FanEn	Digital out	N4	Active high, DC motor drive enable
RC5	FanInt	Digital in	CCP1	DC motor pulse feedback OPT. INT
RD1	Heat1	Digital out	N1	Active high, heater 1 on, DRIVE 1
RD2	Heat2	Digital out	N2	Active high, heater 2 on, DRIVE 2

loads are controlled at the N drive inputs, which operate single-ended in sink mode, since the current drive is needed in only one direction. The P gates can remain disabled. The fan speed is controlled using a CCP module in capture mode. This allows low speeds to be measured accurately.

The temperature sensor is calibrated at 10 mV/°C, with an operating accuracy of ±2°C and offset of 500 mV at 0°C. The temperature range is 0–50°C, so the sensing range is 500–1000 mV. If the second pot is used to provide a reference voltage of 1.024 V, the 10-bit conversion factor is 1 mV per bit, and the temperature is easily calculated in the program by subtracting 500 from the input.

The light sensor needs to be tested to establish the output level when exposed to sunlight and a threshold value incorporated into the program, so that the cooling boost cuts in at an appropriate level. When testing the system, hot and cold air could be applied to the temperature sensor to check basic functionality, but the set temperature input provides a more convenient test input. If the temperature at the sensor is constant (room temperature), adjusting the set input above and below this value has the same effect as the temperature falling and rising.

If the application functions correctly, when the set temperature is adjusted to the actual room temperature, neither the heater nor motor output is on. If the set value is increased, meaning the input temperature is too low, one heater comes on. If increased further, the other heater comes on. If the set value is decreased, the input appears too high and the fan comes on. As the set value is further decreased, the fan speeds up. When the set value is returned to room temperature, all outputs are disabled. If either sensor input is disconnected (the most likely fault mode), the fault output comes on and all other outputs are disabled. The same effect is observed if the motor is stalled, simulating a fan fault.

When the real system is commissioned, the program values may need to be adjusted to optimize the system response. In this kind of feedback system, the system generally needs to respond as quickly as possible without showing instability. The loop delay time (wait for fan) and the PWM calculation might need to be modified accordingly. In commercial temperature controllers, time constant and gain values are adjustable, so that the system response can be optimized in situ.

Listing 4.16 outlines the temperature controller program.

Listing 4.16   Temperature Controller Program Outline

```
TEMCON temperature control system

  Define & Initialize
    StartIn = RA5       (0/1)          Heat1  = RD1 (on/off)
    StopIn  = RA6       (0/1)          Heat2  = RD2 (on/off)
    LightIn = RA1       (0-255)        FanPWM = RD6 (0-255)
    TempIn  = RA0       (0-255)        FanInt = RC5 (0-255)
    SetTemp = RA3       (0-255)        Fault  = RD5 (0/1)
    RunOK   = RD4       (0/1)          Reset  = RE3 (0/1)
    Sunlit  = 0-255     (calibrate)    FanEn  = RD7 (0/1)

  Startup
    All outputs disabled
    Loop
      Read, store, display SetTemp
    While Start button not pressed

  Main Loop
    Read InputTemp

    If InputTemp out of range
      Disable outputs
      Wait for reset
      Flash fault indicator

    If (TempIn-SetTemp<(-2))
      Switch on Heat1
      Disable Fan

    If (TempIn-SetTemp<(-5))
      Switch on Heat2
      Flash fault indicator

    If (TempIn-SetTemp>1))
      Read FanInt
      Calculate fan speed
      Calculate PWM duty cycle

      Read LightIn
      If LightIn out of range
        Indicate fault
        Disable outputs
        Wait for reset

      If (LightIn>Sunlit)
        Add offset to PWM duty cycle
      Modify FanPWM duty cycle
```

```
      Enable fan
      Disable Heaters
      Wait 5s for fan to start
      If (speed=0)
        Indicate fault
        Disable outputs
        Wait for reset

   Else enable RunOK
  Always
```

4.7   PICDEM Board Simulation

- Mechatronics board simulation schematic

- Mechatronics board circuit operation

- Mechatronics board applications

A simulation version of the PICDEM mechatronics board created in Proteus VSM is provided on the support Web site www.picmicros.org.uk. The ISIS schematic is shown in Figure 4.14. The circuit has been organized into functional blocks, and some hardware features are not included to simplify the schematic.

For example, generic drive FETs were used for compactness on the schematic, rather than the specific devices. It was not necessary to include the circuit of the optical interrupter interface, since the DC motor and pulse encoder are modeled in VSM as one component. The RS232 interface is designed to work primarily with a terminal software module provided with the PICDEM kit and therefore also was not included. Components such as decoupling and filtering are used only where essential for accurate circuit modeling. The overcurrent sensing circuit has a simulated input added because variations in the motor loading cannot be represented; this also allows the operation of this part of the circuit to be tested independently. The back emf from the DC motor can be modeled by a voltage source or simple pot if required.

The component numbering is the same as the hardware wherever possible. The circuit connections between the main blocks are made via terminal labeling in the schematic. User connections for particular applications can be added as required.

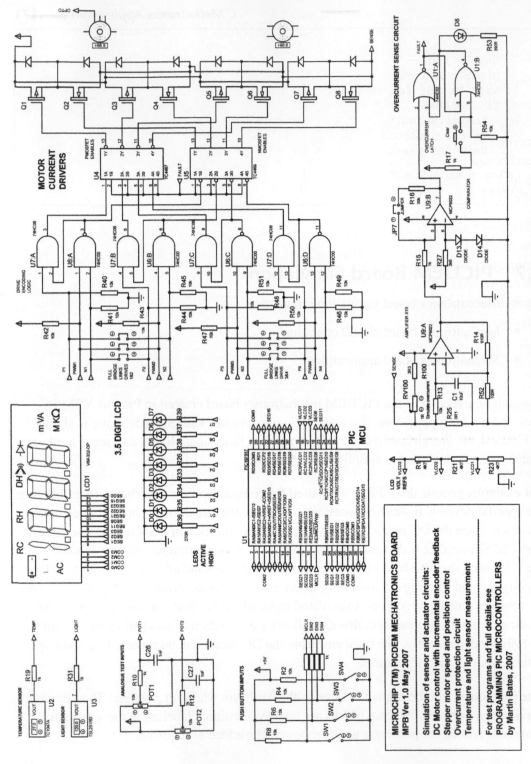

Figure 4.14: PICDEM Mechatronics Board Simulation Schematic

Circuit Description

The central component of the PICDEM mechatronics board is the PIC 16F917, whose main distinguishing feature is the integral LCD drive facility. The 3.5-digit LCD outputs occupy a large proportion of the available I/O pins, leaving a limited number for the other peripherals. The digit segments are enabled by appropriate combinations of the segment and common inputs (see Section 4.2 for details). These are defined in an include file, which must be added to the application project. Three bias voltages are also required by the LCD at V_{cc}, $2V_{cc}/3$, and $V_{cc}/3$; these are generated by a simple resistive divider.

The push-button (tactile switch) inputs on the hardware are represented by toggle switches, so that they can be left in the closed position if necessary when running the simulation. They can be replaced with buttons if preferred. A bank of active high LEDs are provided for output monitoring. The temperature and light sensors are modeled as generic devices, with user control of the set variable. They normally are connected to an analog input on the MCU, either a comparator or an ADC input.

The drive control logic is also modeled using generic devices for the discrete CMOS gates but with specific devices for the enable logic. The driver MOSFETs themselves are generic, so actual device characteristics may not be represented exactly. This is not a significant issue, since the motor models are also generic.

The PMOSFET is switched on when its gate is taken low, and the NMOSFET is switched on when its gate is logic high. No additional interfacing is necessary, which is a great advantage of the FET over other types of current driver, such as bipolar power transistors. In addition, the FET is voltage operated and input resistance at the gate is very high, giving negligible loading on the control logic outputs.

The flywheel diodes in the output are added to cut off the back emf from the inductive motor load when switching off the windings, a standard arrangement with inductive loads. This high-voltage pulse could otherwise damage the FETs. The specific FETs used in the actual hardware have Schottky diodes across the outputs, which perform a similar protection function.

A motor overcurrent is detected by a $0.1\text{-}\Omega$ resistor, through which all driver currents flow to the ground. This generates a voltage of 100 mV at 1 A, and a noninverting amplifier with a gain of 10 increases this to 1.0 V. This voltage is monitored by a comparator stage, which has a reference voltage generated by a pair of diodes in series giving just over 1 V. When this voltage is exceeded, the comparator output triggers the overcurrent latch, which disables the bridge drivers via their control logic. This latch needs to be reset via the CLR FAULT push button on power-up or when an overcurrent condition has been cleared.

Logic functions controlling each half bridge driver have been derived from inspection of the control logic in the schematics of the mechatronics board in the *PICDEM User Manual*.

Source FET on: `!Pg = P.F.(!(M.N))`
Sink FET on: `Ng = M.N.F`

where

Pg = PMOSFET gate (active low),
Ng = NMOSFET gate (active high),
N = N input from MCU,
P = P input from MCU,
M = PWM input from MCU,
F = FAULT input (disable all outputs).

The operation of each bridge driver deduced from these functions is represented in Table 4.8, which shows only the significant logic conditions. The full logic table confirms that the important fact that the FETs are never on at the same time, which would effectively short out the drive supply. F always disables the output when low (power-up condition from the overcurrent circuit). For most input combinations, the half bridge is disabled (safe).

When the bridge control inputs are not connected, the P and N inputs are pulled low (0), the M input pulled high (1) (logic states shown in bold), and the outputs are disabled (Pg = 1, Ng = 0, State 2). They are also unconditionally disabled when F is low (Fault mode, State 1).

Table 4.8: Bridge Driver Control Logic States

Inputs				Outputs		Result	Drive	State
P	M	N	F	Pg	Ng			
X	X	X	0	1	0	Bridge disabled, both off	OFF	1
0	X	**0**	1	1	0	Bridge disabled (default input)	OFF	2
1	1	0	1	0	0	Source on, Sink off	SOURCE	3
X	1	1	1	1	1	Source off, Sink on	SINK	4
1	0	X	1	0	0	Source on, Sink off	SOURCE	5
0	0	X	1	1	0	Bridge disabled, both off	OFF	6
Note: Default input (open circuit links) is shown in bold.								

Assuming we start with all inputs open circuit and both FETs off, the bridge is switched to the Source mode when the P input is taken high (State 3) and to the Sink mode when N is taken high (State 4). The Sink mode can be used to switch a load connected to the positive supply on and off or to provide single-ended PWM drive.

For full-bridge operation, P1 and N2, P2 and N1, and M1 and M2 are linked via the six input links. Drive 3 and 4 inputs are linked in the same way. In this mode, load current is bidirectional and can be reversed by toggling M with P and N high (States 4 and 5). States 4, 5, and 6 allow the bridge to be switched between Sink, Source, and Off.

Demo Applications

The mechatronics board simulation represents fixed connections around the MCU by labeled terminals. Additional connections can be made to uncommitted pins using the normal wiring tools in ISIS, allowing the demo applications to be tested. Note, however, that only the full version of ISIS is guaranteed to allow complete control of the simulation. Therefore, different versions of the mechatronics board schematic configured for testing particular applications are provided on the support Web site.

Assessment 4
5 points each, total 100

1. Sketch a full bridge driver circuit using PFETs and NFETs connected to a motor, indicating the current flow for forward motion and the logic state of the FET inputs.

2. Calculate the speed of the stepper motor on the mechatronics board in rev/min if it is driven at a rate of six steps per second.

3. Derive a formula for the output of the temperature sensor on the mechatronics board, in the form $V = f(t)$.

4. Suggest three disadvantages of using the 3.5-digit parallel LCD compared with the serial alphanumeric display described in Part 2.

5. Write a statement to display the number '8' on digit 1 on the mechatronics board LCD, and explain the meaning of each element of the statement.

6. Describe briefly the hardware used to control the speed of a DC motor connected to a microcontroller.

7. Outline how the position of the stepper motor on the mechatronics board is controlled and the connections required.

8. Outline a method for controlling the speed of the DC motor in the mechatronics board, using Timer1 in the MCU to measure the sensor pulse period.

9. Calculate the delay required in the STEPTEST Program to run the stepper motor at about 1 rev/sec (full step mode).

10. The temperature sensor on the mechatronics board has a calibrated output, while the light sensor does not. Explain why the comparator interface is therefore appropriate for light sensing but the ADC would be preferred for temperature measurement.

11. The temperature at the mechatronics board sensor is 25°C and is converted by the 10-bit ADC with a reference voltage of 2.048 V. Calculate the ADC output value.

12. Write down logic functions for the Source (Pg.Ng) and Sink (!Pg.!Ng) conditions of the board driver logic in terms of the input variables P, M, N, and F from the logic states shown in Table 4.8.

13. List the hardware links required for the bidirectional DC motor drive in the mechatronics board, and explain their significance in terms of switching the current in the bridge forward, reverse, and off.

14. State the connections required for the stepper motor drive in the mechatronics board, and list the activation sequence required at the drive logic inputs.

15. State the features of the power MOSFET that make it suitable for use as a current driver device.

16. Refer to the simulation schematic Figure 4.14 and calculate the output voltage of the overcurrent amplifier in the mechatronics board simulation circuit when the test pot is set to its mid-position.

17. Refer to the simulation schematic Figure 4.14 and explain briefly how the overcurrent latch functions.

18. Explain briefly why a PMOSFET and an NMOSFET are needed in each half-bridge driver stage.

19. Outline how to set up the mechatronics board to control the speed of the DC motor in one direction only, and state the required output from the MCU.

20. Study the setup for stepper motor driving in full-bridge mode; and by using the drive logic functions, determine the winding activation sequence, in terms of the current flow between drive terminals 1, 2, 3, and 4.

Assignments 4

To undertake these assignments, install Microchip MPLAB (www.microchip.com), Labcenter ISIS Lite (www.proteuslite.com), and CCS C Lite (www.ccsinfo.com). Application files may be downloaded from www.picmicros.org.uk. Run the applications in MPLAB with Proteus VSM selected as the debug tool. Display the animated schematic in VSM viewer, with the application COF file attached to the MCU (see the appendices for details).

Assignment 4.1

Download the mechatronics board simulation file `PICDEMboard.DSN` and attach the program `test.cof`. Check that the simulation runs correctly, causing the outputs at Port D to display a binary count. Modify the delay count and confirm that the output timing changes accordingly.

Assignment 4.2

Download the PICDEM mechatronics board simulation file `PICDEMdcmotor.DSN` and attach the program `motorsim.cof`. Check that it runs correctly, displaying the motor revs completed on the display after the input switch has been operated. Modify the program to measure the time interval between pulses from the motor sensor and display the speed in rev/sec. To implement this, measure the pulse interval in microseconds using Timer1 (maximum count = 65 ms) in Capture mode, MCU clock = 4 MHz. This gives the time taken for half a rev in microseconds, t_h, and the speed can then be calculated, in rev/sec = $10^6/2t_h$. For example, if the speed is 3000 rpm (probably exceeding the maximum achievable by the motor), we should see 50 rev/sec on the display. The value of t_h will then be 10 ms, a count of 10,000 in Timer1. We can see from this that the minimum speed measurable is about 10 rev/sec. Use a suitable prescale value to extend this value to less than 1 rev/sec, and modify the program to improve the precision of the speed measurement to ± 0.1 rev/sec.

Assignment 4.3

A temperature controller program is required for the mechatronics board that implements a cooling system. The DC motor has a fan attached, and the controller increases the fan speed when the temperature increases. Connect up the mechatronics board for PWM control of the DC motor. Write a cooling program that reads the temperature sensor and modifies the motor speed accordingly. Demonstrate the application in simulation or hardware as facilities allow.

PART 5

PIC16 C Applications and Systems

5.1 PIC16 C Application Design

- Block diagram

- Program outline

- Debugging and testing

Formal design methods recommended for engineering projects may need to be applied in the professional design environment. Here, some basic methods are outlined as a starting point; these allow new applications to be developed with some degree of consistency and help communicate project concepts and design details clearly in reports and presentations.

Hardware Design

The block diagram is an effective way to show the general form of a microcontroller application design, and examples are seen throughout this book. Some simple rules are used to represent system blocks and their input and output signals:

- The direction of signal flow is represented by an arrowhead.

- The TTL level digital signal is the default (default arrow) style.

- Other switching levels (e.g., RS232 line) are indicated by labels.

- The analog voltage range is indicated by a label and arrow style.

- Parallel data are represented by a block arrow.

- Analog signals are represented as a simple waveform.

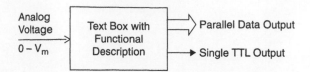

Figure 5.1: Block Diagram Conventions

The block diagram (Figure 5.1) is easily constructed using only the drawing tools in a standard word processor. The example in Figure 5.1 might represent an analog-to-digital converter chip, with an "end of conversion" output.

Once a block diagram has been created, defining the inputs and outputs of each block, a circuit schematic can be derived from it.

Software Design

The application program can be designed using various methods. A flowchart shows the overall program sequence in a visual manner and is good for illustrating simple program sequences. However, for C programs, some form of structured pseudocode is recommended, where the main program is outlined as a text file, which can then be converted directly to source code. Examples are again found throughout this book; the general content is described in Listing 5.1.

After the application program source code has been created in the MPLAB text editor, it can be compiled to generate the project file set. This includes the MCU machine code HEX file and the COF file, which incorporates the hex file with additional debugging information. It is necessary to have all the project files in the same folder, making copies of the resource files as necessary. All applications need an MCU header file, such as 16F877A.H.

The application source code, MCU header file and any other files to be included or used must be attached to the project in the project file window. The application can then be built and the HEX machine code file produced. This is downloaded to the target system to operate the application in hardware.

Application Debugging and Testing

The application program is tested and debugged in several stages. The main types of errors and the tools for detecting them are outlined next.

Listing 5.1   General Control Program Outline

```
PROGNAME.C /////////////////////////////////////
  Program header information
  Author, date, version etc

  Include MCU header file
  Include function library files
  Include user source files
  Use function library files
  Define constants

  Declare global variables
  Declare function prototypes

Main block //////////////////////////////////////
  Initialization sequence
  Initialization function calls
  etc

  Main loop
    Sequences
    Function calls (level 1)
    etc and repeat

Function block //////////////////////////////////
  Initialization sequence
  Process sequences
  Function calls (level 2)
  etc and return
```

Syntax errors are mistakes in the source code, such as spelling and punctuation errors, incorrect labels, and so on, which cause an error message to be generated by the compiler. These appear in a separate error window, with the error type and line number indicated so that it can be corrected in the edit window.

When the program is successfully compiled, it can be tested for correct function in the target hardware so that any logical errors can be identified. However, it is preferable to test it in software simulation mode first, as it is quicker and easier to identify errors in the program sequence. Two simulation methods are available here, MPSIM and Proteus VSM.

MPSIM is the simulator provided with MPLAB. It allows the program source code to be run, stopped and stepped, and breakpoints set. The registers and source variables may be inspected at each step. When debugging C programs, breakpoints are the most useful, while stepping is more useful in assembly language. The program sequence and variable values are monitored and errors identified when the results obtained do not agree with those expected. Error information is provided principally in tabular form.

By comparison, the Proteus VSM debugging environment has significant advantages. The animated schematic gives a much more immediate indication of the overall program function. Interactive input and output devices operate in real or simulated time. The source code and breakpoints can be displayed.

In addition, if the VSM viewer is run from within MPLAB, the progress of the program can also be monitored simultaneously in MPSIM. Therefore, the more detailed debugging tools in MPSIM can be run alongside VSM and the most appropriate selected for any debugging task. The simulated hardware design is thus tested in conjunction with the MCU firmware (cosimulation), allowing circuit modifications at an early stage and hardware-software interaction to be studied on screen. When the program is eventually downloaded to the real hardware, it is now far more likely that it will work the first time.

The VSM Viewer is invoked from the debug tools menu in MPLAB, and the program is attached and tested. However, if circuit modifications are needed, VSM must be opened separately to run alongside MPLAB, so that the full set of ISIS schematic edit tools and component models are available. VSM still accesses the same COF file, so both software and hardware changes can be tested. More details on interactive debugging are given in Appendices A, B, and C.

5.2   PIC16 C Temperature Controller

- Basic system
- Software design
- Implementation

In this section, the software design principles just outlined are applied to a typical application, a temperature control system. The schematic of the demo hardware is shown in Figure 5.2. The TEMP pot represents a temperature sensor that outputs a voltage of 0–5 V. If a scaling of 100 mV/°C is assumed, the range is 0–50°C, with 2.5 V representing 25°C.

System Operation

The sensor is connected to AN1, the ADC channel 1. A SET pot provides the reference temperature for the system. If the measured temperature is below the set level, a heater,

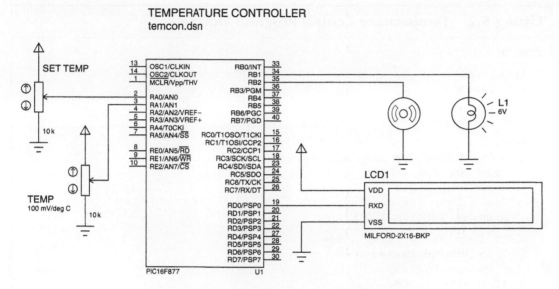

Figure 5.2: Temperature Control System

represented by the filament lamp output, is switched on. If it is above the set value, a cooling fan switches on instead, represented by the DC motor.

To avoid the outputs "chattering" at the switching point, due to input noise, switching hysteresis should be incorporated into the control sequence, meaning that the switching level when the temperature is rising is higher than when the temperature is falling.

The temperature is displayed on the serial LCD as well as the status messages Heater ON or Fan ON. The program structure ensures that the correct message is displayed during the changeover phase.

Software Design and Implementation

The process of designing the software can be aided by writing a program outline. The main structures and sequences are summarized using suitable layout and operational descriptions.

A typical problem to be overcome is that the displayed message must agree with the output status in the presence of hysteresis. Therefore, an output status flag (variable type int1) is used to record the current output status. This flag is then tested by the conditional output statement. Note that the switching levels can be modified to suit the application. In the code shown (Listings 5.2 and 5.3), the upper switching level is 20 steps above the lower.

Listing 5.2   Temperature Control Program Outline

```
TEMCON

Initialize
  MCU 16F877A
  ADC 8 bits, Inputs RA0, RA1
  RS232, Output RD0

  Loop
    Delay 500ms for display
    Read Set Pot 0-255
    Read Temp 0-255
    Scale Temp for display
    Display Temp on LCD line 1

    If Temp below lower limit
       Switch ON Heater
       Switch OFF Fan
    If Heater is ON
       Display on LCD line 2

    If Temp above upper limit
       Switch OFF Heater
       Switch ON Fan
    If Fan is ON
       Display on LCD line 2
  Always
```

In a real system, the interfacing needs to be further developed. The temperature sensor is likely to need an amplifier, perhaps with voltage-level shifting. The heater and fan need a relay or contactor to operate the final load, with the relay requiring a transistor interface or current driver. Details of interface design can be found in *Interfacing PIC Microcontrollers* by the author.

5.3   PIC16 C Data Logger System

- BASE board hardware

- Application design

- Program outline

Since this book is concerned mainly with software development, off-the-shelf hardware, such as the PICDEM mechatronics board featured in Part 4, is very useful. This is

Listing 5.3   Temperature Controller Source Code

```
/*
  TEMCON.C MPB 27-3-07
  Temperature controller demo. Target simulation system: TEMCON.DSN

*************************************************************************/
#include "16F877A.h"
#device ADC=8                               // 8-bit conversion

#use delay(clock=4000000)
#use rs232(baud=9600, xmit=PIN_D0, rcv=PIN_D1) // Display output

void main() //*****************************************************
{
  float refin, numin, temp;
  int1 flag;

  setup_adc(ADC_CLOCK_INTERNAL);            // Setup ADC
  setup_adc_ports(ALL_ANALOG);

  for(;;)  // Repeat always
  {
    delay_ms(500);
    set_adc_channel(0);                     // Read ref. volts
    refin = read_adc();
    set_adc_channel(1);                     // Read temp. volts
    numin = read_adc();

    temp = (numin*50)/256;                  // Calc. temperature
    putc(254); putc(1); delay_ms(10);
    printf(" Temp = %3.0g ",temp);          // Display temp.
    putc(254); putc(192); delay_ms(10);

    if (numin<(refin-10))                   // Temp. too low
    { output_high(PIN_B1);                  // Heater on
      output_low(PIN_B2);                   // Fan off
      flag = 1;
    }
    if (flag==1) printf(" Heater ON ");     // Status message

    if (numin>(refin+10))                   // Temp. too high
    { output_low(PIN_B1);                   // Heater off
      output_high(PIN_B2);                  // Fan on
      flag = 0;
    }
    if (flag==0) printf(" Fan ON ");        // Status message
  }
}
```

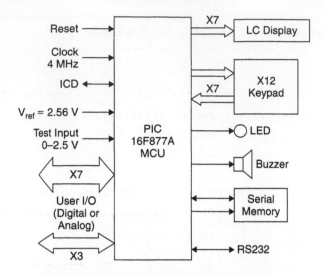

Figure 5.3: BASE Board Block Diagram

reflected in real applications by the use of standard hardware such as PC-compatible boards as the platform for a wide range of applications.

BASE Board

A general purpose board with a typical selection of peripherals attached to a PIC 16F877A is described here. This design was originally developed to demonstrate hardware interfacing techniques. The PIC 16F877 BASE (basic application and system evaluation) board incorporates six analog inputs, a 12-button keypad, a parallel 16 × 2 character LCD, 16 k serial memory, an RS232 port, and ICD programming connections. The block diagram is shown in Figure 5.3, the schematic in Figure 5.4.

Here, the board is used as a data logger. It records input analog voltage levels at timed intervals and stores this data for later uploading to a host PC. The PIC 16F877 has eight 10-bit analog inputs, but to keep the demo system simple, 8-bit conversion is used. The reference voltage applied to RA3 is 2.56 V, which gives a resolution of $2.56/256 = 10\,\text{mV}$ per bit and a precision of $100/256 \approx 0.4\%$.

The reference voltage and a test input occupy two of the analog inputs, so six are available for connecting to an external target system. Typically, the inputs are connected to analog sensor inputs, measuring temperature, position, strain, and other physical variables from suitable sensors. Another possibility is that the target system is an analog board whose performance is being evaluated by measuring the circuit voltages under test conditions.

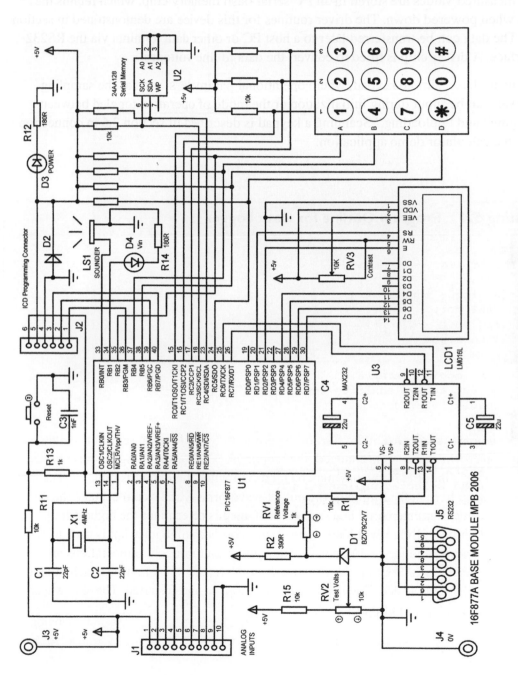

Figure 5.4: BASE Board Circuit Diagram

The measured values are stored in an I^2C serial flash memory chip, which retains the data when powered down. The driver routines for this device are demonstrated in section 3.6. The data can be transferred later to a host PC or other data terminal via the RS232 interface. A driver chip is fitted to convert the data to line voltages.

The board has a simple keypad, where operational parameters, such as the sampling interval, can be input during initialization or the mode of operation toggled between "logging" and "uploading." Scanning a keypad is described in section 2.6 in connection with the calculator demo application.

Listing 5.4   Program Outline for Data Logger

```
LOGGER
  Initialize
    Delays
    Analogue inputs
    UART port
    I2C port
    Interrupts

  Main
    Set logging interval
    Select active analogue inputs
    Enable interrupts
    Wait

  Interrupt Routines
    Timeout
      Restart timer
      Read selected analogue inputs
      Store in external EEPROM
      Display channels and input voltages
      Return from interrupt

    Zerokey
      Disable timer
      Display 'Logging Stopped'
      If Starkey
        Restart logging
      If Hashkey
        Send data via RS232
        Display 'Sending data'
        Return from interrupt
```

The parallel LCD is used to display status messages and data as they are sampled. It is useful to compare it with the serial LCD described previously, as parallel access is generally faster, particularly when bit maps are used for graphics in more sophisticated applications. The 8-bit ASCII and control codes must be sent as 4-bit nibbles from RD4-7, with RD1 acting as the register select (RS) input and RD2 generating the data strobe (E). More details are provided on driving the parallel LCD in *Interfacing PIC Microcontrollers*, by the author. Alternatively, the manufacturer's data sheet can be consulted for the necessary control codes and timing information.

Program Outline

As can be seen in the program outline (Listing 5.4), the application is largely interrupt driven. The timer interrupt is the simplest way to generate a regular event, in this case, sampling at fixed intervals. The 0 key is used to interrupt the logging process, so it might be desirable to reassign the input from column 2 of the keypad to RB0, the primary interrupt input. Logging is restarted using the star (∗) key and data upload initiated using the hash (#) key.

5.4 PIC16 C Operating Systems

- Polling

- Interrupts

- RTOS

As microcontroller operating programs become more complex, consideration must be given to the best method of organizing the program response to input, memory management, and output timing. Three main methods are used to handle input and output events, which after all, is the primary requirement of a real-time system. In order of complexity, they are I/O polling, interrupts, and the real-time operating system (RTOS).

Polled I/O

This is the easiest, and may be considered the default, method of input and output, where operations are simply scheduled as part of the main loop. It is seen in most of the examples in this book, because they have been deliberately kept simple. The basic principle is illustrated in Figure 5.5.

This option is fine if the delay that occurs between input signal and output response is not critical to the correct overall operation of the system. The time taken to complete

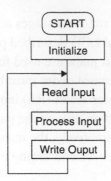

Figure 5.5: Polled I/O Process

the input processing may vary significantly, depending on the input data or programmed options within the loop. For example, a test on the data may result in an optional sequence being executed, or not, depending on the value. In fact, this is pretty much inevitable in most real programs.

However, it is often important for the input and output timing to be more predictable. Take the example of motor speed control. In small DC motors, this is usually implemented by pulse width modulation, as discussed in section 4.3. The output is switched on and off over a regular cycle, the proportion of "on" to "off" time determining the average motor current and hence the speed. To achieve accurate control, the shaft speed must be measured, usually by a pulse encoder. The input pulse interval must be measured and the PWM duty cycle adjusted accordingly. It is just about possible to do this using a polling process (see *PIC Microcontrollers, An Introduction to Microelectronics* by the author, 2004), but a more elegant solution can be implemented using interrupts.

Interrupts

As we have seen in Section 2.9, interrupts are internally or externally generated asynchronous hardware signals that force the processor to stop its current (background) task and carry out the interrupt service routine (ISR), a higher-priority (foreground) task. The processor "context" (current register contents and status) must be saved and the current program address stored on the stack so that the background task can be resumed when the ISR has finished.

Let us see how this can be applied to the motor controller, assuming we are using a 16F877 MCU (Figure 5.6). The input pulse period can be measured using one of the hardware timers. Since Timer2 is designed to provide PWM mode, Timer1 can be used to monitor

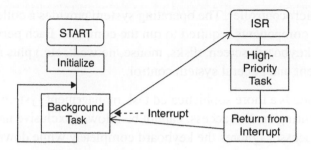

Figure 5.6: Basic Interrupt Operation

the input, working in Capture mode. The counter/timer register is fed from the system clock to measure absolute time intervals, and the count is stored when the input changes.

The pulse period can then be worked out and this result compared with a target value, which represents the required period (hence speed). If it is too long (speed too low), the motor speed is increased by increasing the PWM duty cycle in Timer2. If too short (speed too high), the duty cycle is reduced. An interrupt is generated by Timer1 when the count is captured; the ISR modifies the output duty cycle as required, and the controller then waits for the next interrupt to occur.

If the program uses multiple interrupts, one ISR may be interrupted by another. The interrupts may need to be assigned an order of priority, so that a less important task does not interrupt a more important one. When the higher-priority ISR is being executed, the lower-priority interrupt can be disabled, or masked, until it is finished. In more complex programs, numerical levels of priority can be assigned, with higher priorities taking precedence. Unfortunately, the 16 series PIC is not well suited to this, as it does not have a built-in priority system, unlike more powerful processors. Further, the different interrupt sources have to be identified explicitly by a user routine.

An operating system (OS) provides an alternative to interrupts as a means of providing a more predictable time response in the microcontroller system but again is typically implemented in the higher-power MCU type, such as the PIC18 or 24 series. Nevertheless, to point the way ahead, the principles are outlined here.

PC Operating System

The most well-known example of an operating system is Microsoft Windows®. Why is this needed in PC-type computers? The answer is simply the complexity of the software compared with a microcontroller. The operating system provides a collection of the numerous program components required to run the computer. Each peripheral interface has its own driver (keyboard, screen, disks, mouse, network, etc.) plus modules for memory management and general system control.

Therefore, the PC needs a more sophisticated task management system. A lengthy process, such as printing or disk access, cannot be allowed exclusive use of the system resources. If the processor ignores the keyboard completely while downloading a large file from the Internet, the user cannot access the system to do something more urgent. In addition, the OS has to be multitasking; that is, it must allow several operations to appear to be running simultaneously, such as allowing you to keep writing while printing. We also want to switch quickly between tasks by keeping more than one window open at a time, which means keeping multiple tasks loaded in memory. For example, while running the examples in this book, we need to have MPLAB and Proteus open at the same time, plus maybe a data PDF and the word processor.

Multitasking is essentially achieved by time slicing. Each apparently concurrent task is allowed to run for a given time interval, say 100 ms, then execution switches to another. Priority can be assigned, so that, for example, one Internet data packet is picked up and stored in memory before the next arrives and overwrites it in the network data buffer. Therefore, the OS is designed so that multiple tasks appear to run smoothly together and with the right priorities.

The PC is essentially a batch processing system; that is, the timing of the major tasks is not critical. If a word-processing task is delayed for a few milliseconds, it is not apparent to the user and not significant in terms of overall system effectiveness. On the other hand, the timing of events in so-called real-time systems must generally be highly predictable. When an input is received, it must be processed and the output generated within a known time frame. The point is obvious if one considers an example such as an aircraft flight control system or automobile engine controller. To manage complex control system software, we may need a real-time operating system.

Real-Time Operating System

The principle of operation of a simple RTOS, as implemented by CCS C, is shown in Figure 5.7. The program is divided into separate tasks, which are executed in turn. A timer interrupt causes the task switching, but interrupts are otherwise *not* used. When a task is suspended, its context (file register state) is saved and restored when it is restarted the next time around. In this way, multiple tasks are executed in rotation and can appear to execute

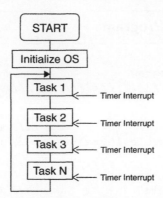

Figure 5.7: Basic RTOS Operation

simultaneously, and the I/O timing is more predictable. More sophisticated systems incorporate task priority and implement more complex task management strategies.

A blank program is shown in Listing 5.5 to illustrate how CCS C implements the RTOS. The MCU used is an 18F452, which is the 18 series equivalent to the 16F877 (CCS C supports RTOS for only 18 series PICs and above). The delays in the RTOS are implemented using the standard function, where the MCU clock rate has to be specified (20 MHz).

The directive #use rtos() indicates to the compiler that this program uses the RTOS structure. It then expects some task definitions to follow and the main block to contain the statement rtos_run(). The hardware timer used to produce the timer interrupt that triggers task switching is specified as an argument of the directive, Timer0 in this case. The "minor cycle" defines the maximum time for which the task runs. Each individual task execution rate must be a multiple of this time.

The task definitions follow. Each is preceded by the directive #task, so that the compiler knows this is an RTOS task and not a standard function definition. The rate specifies how often the task executes (e.g., once per second for Task 1), and max is the maximum time allowed for this task. The task block is then defined as a sequence of statements in the same way as a standard function, but bear in mind that its execution can be suspended and restarted at intervals defined by the RTOS.

All that remains then is to start up the RTOS in the main block, and the tasks are executed in turn, with the frequency and duration specified for each. The CCS implementation is classified as a cooperative, multitasking RTOS. This means that the tasks return control to the scheduler voluntarily to allow the next to run. A set of functions

Listing 5.5   Blank RTOS Program

```
// RTOS1.C
// Minimal blank RTOS program
//////////////////////////////////////////////////////////////////////

#include <18F452.h>                          // Define MCU
#use delay(clock=20000000)                   // Define clock rate
#use rtos(timer=0,minor_cycle=100ms)         // Define RTOS timing

// Task functions /////////////////////////////////////////////////////

#task(rate=1000ms,max=100ms)                 // Define first task
  void task1()
  {
    // Task1 statements...
  }

#task(rate=500ms,max=100ms)                  // Define another task
  void task2()
  {
    // Task2 statements...
  }

#task(rate=100ms,max=100ms)                  // Define last task
  void task3()
  {
    // Task3 statements...
  }

// Main function //////////////////////////////////////////////////////
void main()
{
  rtos_run();                                // Start RTOS scheduler
}
```

are supplied that allow the tasks to work together for optimum effect. For example, rtos_enable(task1) and rtos_disable(task1) allow tasks to be selectively enabled and disabled. The function rtos_yield() allows the task to return control to the scheduler when finished. Some functions allow status information and messages to passed between tasks and the progress of the tasks to be monitored.

The RTOS is implemented with a total of only 13 functions and directives (see the *CCS C Compiler Reference Manual*). A good general explanation of RTOS principles and types can be found in the *Salvo RTOS User Manual*, Chapter 2, from Pumpkin Inc. (www.pumpkininc.com).

5.5 PIC16 C System Design

- Hardware selection

- Software design

- System Integration

We have seen how to get started with building PIC microcontroller systems programmed in C. Simple examples have been used to illustrate the basic principles, so we now need to look at some issues relating to more complex microcontroller-based systems. Numerous texts are available, written by experienced and knowledgeable engineers, that discuss the finer points of real-time system design, so the intention here is to introduce the some basic concepts to help the reader to move toward a further understanding of real industrial applications. Another objective of this section is to review some relevant factors in the selection of the best combination of hardware, programming language, and development tools for any given microcontroller product design.

Hardware Selection

There is a range of related devices around which embedded systems may be designed, including a

- Microcontroller (MCU)

- Microprocessor (CPU)

- System on a chip (SoC)

The conventional microprocessor system embodies the traditional approach, where a central processing unit, memory, and peripherals can be put together to meet the requirements of a particular application as precisely as possible. Designing a custom-made CPU system is a relatively expensive option, and such an extensive range of other options are available that the conventional CPU-based system may be needed for only highly complex, specialist systems or where a low-cost, standard board such as the PC motherboard can be easily adapted. The discrete microprocessor does, however, allow multiprocessor systems to be designed that typically use shared hardware resources, especially memory. Current standard processors typically incorporate features to support multiprocessor operation, and the dual core processor is currently becoming standard in PCs.

The SoC takes the concept of the microcontroller to the next level. It is, in effect, a configurable microcontroller, where the designer has control over the internal arrangement of the hardware elements. Using a dedicated design system, the processor core is selected and the required memory and peripherals added. These hardware elements are supported by corresponding standard drivers provided as part of the package. With a complex interface, such as USB, for example, the provision of a standard protocol stack (software layers, not a hardware stack) is essential. The design can be fully tested in software, in the same way that a PIC program can be tested in MPLAB. Only when finally verified is the design fabricated by the hardware supplier.

If a design is to be created from scratch, then the most appropriate type of system may be selected from the three main options listed previously. However, this choice is unlikely to occur in isolation; factors such as the previous experience of the design team, existing company products, and so on are significant. Nevertheless, the designer should keep an open mind as far as possible and needs to keep up with a rapidly developing technology in the embedded systems field to make the right choice—not easy.

Microcontrollers

A designer who has a store of expertise using a particular microcontroller type and development system will need a good reason to look elsewhere for a solution. Gaining similar expertise in another system takes time and resources, and any change must also take into account the future strategy of the company or design group.

The PIC family may be our first choice for the following reasons:

- Low cost

- Simplicity

- Good documentation

- An established market

- A development system provided

- Third party support

The PIC is well suited to the learning environment as it was originally pitched at the low-end (high-volume, low-complexity) market and is well supported by third party products. Therefore, the assumption implicit in this book is that the PIC is the best starting point, even if the learner is later to progress to other processor types. At the time of this writing the main alternatives are Atmel (AVR), Freescale (Motorola), STMicroelectronics, Hitachi, Philips, and National Semiconductor.

We can approach hardware selection on the basis of the choice offered within the PIC range, which was outlined in section 1.1. Some of the main features to consider are

- The number of I/O pins.

- The interface types.

- The program memory capacity.

- The RAM capacity.

- The operating speed.

- The power consumption.

We assume that adequate development system support and driver libraries are available. A logical approach to design is to select a chip that has spare capacity in relation to the draft specification. The application can be prototyped in simulation mode without penalty using an overspecified device. When the I/O, memory, and peripheral requirements finally are established, a chip can be selected for hardware implementation that meets the specification at minimum cost.

The anticipated scale of production is also a factor. The cost of each individual unit produced becomes more critical as the scale of production increases. On the other hand, the firmware can be reproduced at effectively no cost, unless variants are required. If we assume a fixed cost, a, for design development (hardware and software) and each board costs b to produce, the cost per unit is given by

$$y = a/x + b$$

where x is the number of units produced. The fixed costs are divided by the number of boards produced. So, if the development costs are, say, 1000 units of currency ($a = 1000$) and the production cost 100 per board ($b = 100$), a curve showing the cost per board as the volume of production is increased is obtained, as seen in Figure 5.8. We can see that the cost per board is initially high, falling away and leveling off as the production volume increases.

Hardware Design

Taking the hardware design criteria in turn, we can consider some of the relevant factors in getting started with a design, assuming an agreed-on initial specification. Having said this, it is useful to know how much flexibility is allowed in meeting the specification,

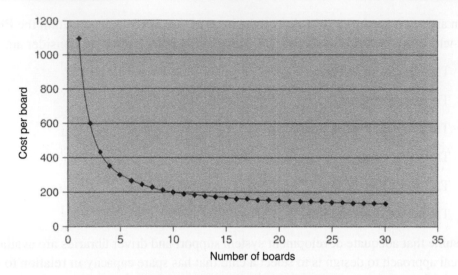

Figure 5.8: Production Cost

because a disproportionate cost might be involved. It may be acceptable to reduce the performance to reduce costs, for example, reducing the precision of analog measurements or the frequency range of a signal output.

The cost of the microcontroller tends to increase with the number of I/O pins, so it is probably a good idea to look for ways to reduce the pin count. One example we saw in previous sections is to use a serial LCD instead of a parallel one. The serial type requires only 1 output, while the parallel LCD seen earlier needs 7, or possibly 11 if 8-bit data are used. Certainly the serial interface should be considered the default choice, and the parallel used only if high-speed access to the display is needed. The serial link can also be physically longer.

Serial access sensors are becoming more common, where the data are sent to the MCU in serial form, rather than as an analog signal. We saw that any pin can be used as an RS232 port, because CCS C provides a driver that generates the required interface purely in software. This means dedicated analog ports may not be necessary, giving greater flexibility in the choice of MCU. On the other hand, the sensor is likely to be more expensive.

Program memory capacity requirements are not easy to anticipate before the software has been finalized. C programs generally need more memory than assembler, so the choice of language is important. This factor is considered further later, but for now, suffice it to say that memory requirements expand rapidly with program complexity. As regards RAM requirements, the PIC is strictly limited, as the only onboard RAM consists of spare file registers. External data memory may well be necessary, as in our data logger. An alternative type of MCU could even be necessary for data-intensive applications.

The PIC scores well on operating speed, however. The 16 series devices can generally run at 20 MHz, with the 18 and 24 series running at 40 MHz. The clock speed does affect the power consumption, as the current consumed is proportional to the switching rate in CMOS devices. Low-power MCUs are an important ongoing development in microprocessor technology. Reduced operating voltages (e.g., 3.3-V supply) are also increasingly used to reduce power dissipation. Power consumption is not one of the operating parameters normally predicted by simulation, so a real hardware prototype may be needed to finally specify the power supply. Obviously, power consumption is even more critical in battery-powered systems.

Software Design

There are two main options for creating the system firmware for low-complexity embedded systems: assembly language or C. There are other user-friendly programming options aimed primarily at learners, such as software that allows C code to be generated from a flowchart (see Appendix D). A wider range of high-level languages and proprietary development systems are available to support more advanced processors.

In general, assembly language is used for simple programs and those where direct access to control registers or speed is critical. Certainly, using assembler requires an intimate knowledge of the MCU architecture and is an essential tool for the practicing embedded engineer. If necessary, assembly language blocks of code can be embedded within a C program.

However, the premise of this book is that there are good arguments for starting with C. Less detailed hardware knowledge is needed, and programming is simplified. It is also a universal language, whereas each MCU type has its own assembly language. Used in conjunction with a user-friendly simulator, such as Proteus, useful applications for any microcontroller type can be created with a minimum of experience. The availability of a comprehensive set of peripheral drivers is also very helpful, as provided by CCS C. However, the main advantage is that C is by far the most widely used high-level language for embedded systems and can be applied by all embedded engineers, from beginner to expert.

The overall structure of the embedded firmware is determined by the complexity and, to some extent, the hardware features of the host MCU. A simple program can use polled

I/O in assembler program. If the chip has an interrupt structure that allows task priority and timing to be adequately managed, then interrupts can be used in assembler or C. The RTOS approach may well be the best solution for more advanced applications; this is the next stage in microcontroller system design, to which I hope the reader will be able to progress because of the system design concepts outlined in this book.

There is never a perfect solution to the embedded design challenge, but we can try for the best one that lies within our own limits of experience and enjoy the challenge it presents.

Assessment 5

5 points each unless otherwise stated, total 100

1. Explain why hysteresis is useful in processing switched inputs.

2. Write two C statements that select analog input AN1 and read it, and explain briefly why the variable comes first in the read statement but is given as the function argument in the select statement.

3. Draw a block diagram of a simple temperature control system, consisting of a temperature sensor, heater, fan, start and stop buttons, and status indicators for "running" and "temperature OK." (10 points)

4. Write a basic program outline for the system described in Question 3 which has a single fixed operating temperature and no hysteresis. A polling loop will wait for the start button to be operated, while the stop button will shut down the system via the MCU reset input.

5. Explain briefly why analog inputs, serial flash ROM, and a serial data link are useful features of data logging system hardware.

6. Explain briefly how the use of a timer interrupt allows an accurate data logging interval to be more easily implemented than simple input polling.

7. Explain briefly the meaning of *interrupt priority*.

8. Compare briefly the different features of a standard PC operating system and an RTOS.

9. Explain briefly the significance of each part of the CCS C RTOS task definition directive #task(rate=500ms,max=100ms).

10. Explain briefly the main difference between a microprocessor and microcontroller-based hardware system.

11. Explain briefly the main advantage of a SoC when compared to a conventional microcontroller.

12. State five criteria for selecting a microcontroller type or family.

13. State five criteria for selecting a microcontroller for a given application.

14. Explain briefly why the cost of a microcontroller application prototype is relatively high, but the cost per unit reduces as more systems are produced using that design, and sketch a curve that illustrates this fact.

15. Compare briefly the merits of a serial alphanumeric LCD module and the DMM display used in the PICDEM mechatronics board.

16. Discuss briefly the factors that affect power consumption in an embedded system and how to evaluate it.

17. Explain the advantages of using C for embedded applications. (10 points)

Assignments 5

To undertake these assignments, install Microchip MPLAB (www.microchip.com), Labcenter ISIS Lite (www.proteuslite.com), and CCS C Lite (www.ccsinfo.com). Application files may be downloaded from www.picmicros.org.uk. Run the applications in MPLAB with Proteus VSM selected as the debug tool. Display the animated schematic in VSM viewer, with the application COF file attached to the MCU (see the appendices for details).

Assignment 5.1

Download the project TEMCON and check that it runs correctly in MPLAB with Proteus VSM viewer. Modify the program to display warning messages when the temperature is more that 3°C above the upper switching level (TOO HOT) or more that 3°C below the lower switching level (TOO COLD).

Assignment 5.2

Design a controller for a small hot and cold drinks machine, aimed at the domestic market. Write a specification based on your own understanding of the typical requirements of such a machine, draw a block diagram showing the interfacing required, and outline a control program which can be implemented in C. Predict the input, output, and memory requirements and select a PIC microcontroller (www.microchip.com) which provides the features required for this application at minimum cost.

Hardware Design Using ISIS Schematic Capture

Proteus VSM is an interactive electronics design package from Labcenter Electronics that allows analog, digital, and microprocessor circuits to be subjected to virtual testing before the creation of a PCB layout for the construction of real hardware. ISIS is the schematic capture package, and ARES is the layout package.

The circuit is entered directly onto the schematic by selecting components from a library of parts, which have associated mathematical models (e.g., V = IR for a resistor). When completed, the wiring schematic is converted to a set of nodes connected by components, represented by a set of simultaneous equations derived from the model for each component. The network is solved for any given set of inputs and the outputs are displayed via active on-screen components, virtual instruments, or charts.

The microcontroller is simulated on the basis of its internal architecture and the specific program being executed, which must be attached to complete the model. In our case, the program is written in C and the COF file produced by the compiler attached to the MCU. This file contains the program machine code and some additional information to help with debugging the program. ISIS allows the source code and variables to be displayed so that the program operation can be studied step by step and any functional errors corrected.

Design Specification

The starting point for an electronics design is a specification, which should state clearly the system performance requirements. Our example project is called BAR1 (Figure A.1).

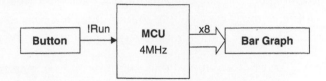

Figure A.1: BAR1 System Block Diagram

This is used as the project folder name and the file name for the project files. The specification is as follows: When a button is pressed, the system generates an 8-bit binary count, starting at 0, on a bar graph display. The output frequency at the least significant bit is 50 Hz, giving an overall cycle time of 2.56 sec.

This specification could be elaborated by, for example, requiring a battery supply. In that case, an LCD display would be preferred for its low-power consumption over the LED display used in the prototype.

A block diagram is useful for clarifying the hardware design. The function of each main circuit block should be identified, as well as the signals in and out. In digital circuits, the polarity of the signal can be indicated (!Run = active low input) and a parallel output represented with a block arrow (x8 = 8 bits). The standard word processor has all the drawing tools needed to create simple block diagrams.

Schematic Circuit

The circuit in Figure A.2 shows a PIC 16F877A with crystal clock, push-button input, and 8-bit bar graph display. The output increments when the button is "pressed" using the mouse pointer, and the effect can be seen on screen in real time.

The design of the circuit obviously requires knowledge of the relevant interfacing techniques to connect up peripheral components correctly. For example, the resistor value in the switch pull-up circuit is not critical, but the maximum value is limited by the input current drawn by the PIC input; a maximum of $1\,M\Omega$ is appropriate. At the low end of the viable range, power conservation is the relevant factor. To limit the current when the switch is closed, a resistor value of at least $1\,k\Omega$ is required; $10\,k\Omega$ is a suitable compromise.

The resistance of each element in the series resistor pack controlling the LED segment currents must be calculated. If the LED current required is assumed to be 10 mA and the

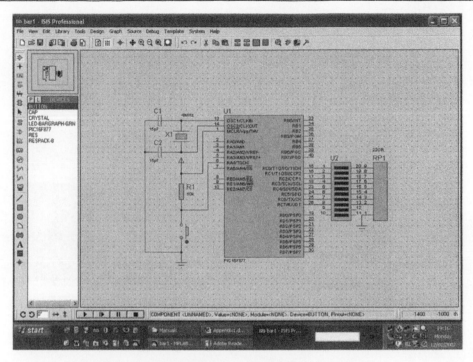

Figure A.2: ISIS Schematic Capture Screen

forward volt drop of the LED is 2 V, then the resistor value is given by R = (5 − 2)/ (10 × 10^{-3}) = 300 Ω (NPV = 270 R).

Refer to *Interfacing PIC Microcontrollers* (Elsevier, 2005) by the author, for further information on interface design.

Schematic Edit

ISIS is opened as a discrete package within the Proteus VSM suite. Create a new design file and save it as BAR1.DSN in a project folder called BAR1, which is accessible from Proteus and MPLAB.

To start the schematic, the Component button should be clicked to enable the Devices mode in the object window. The Pick Devices button [P] at the head of the Object Selector panel gives access to the device libraries (Figure A.3). The category Microprocessor ICs has a subcategory, PIC 16 Family, from which the PIC 16F877 can be selected; it then appears in the device list.

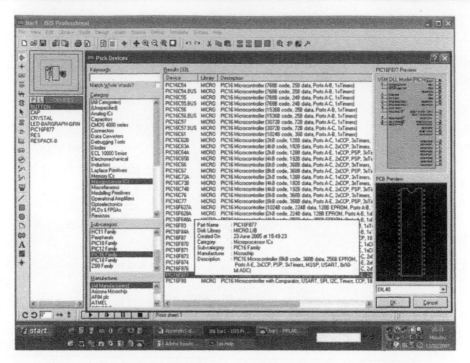

Figure A.3: Picking the Microcontroller from the Parts Library

The bar graph component is picked from the Optoelectronics category, the crystal from Miscellaneous, and the push button from the Switches and Relays. The resistor and capacitor are the generic type. ACTIVE components with an associated SPICE model must be used for interactive testing. Not all components are active, just a representative selection.

After selection from the object list, a component can be placed with a left click on the schematic, highlighted (red) with a right click, and removed by right clicking again. Components are connected together by clicking on the pins in the Component mode. Wires can be connected, but space on the connecting wire must be allowed between pins. Always connect in line with a pin and check that a dot appears to confirm that a junction between pins has been created. The Terminal button brings up the TERMINAL list. The Ground and Power pins can then be placed. The Power pin automatically adopts the V_{dd} of the MCU (+5V).

The Overview window allows the schematic to be recentered and displays the components. The schematic can also be zoomed and centered using the mouse wheel.

Components can be oriented or flipped using the rotation and reflection buttons, and groups of selected components moved or copied using the Tagged Object edit buttons. Each editing feature should be explored by reference to the Proteus help files and practical experiment.

The clock circuit and power supplies are implicit in the microcontroller model, so it is not actually necessary to include the external clock components at this stage. However, they must be added before a circuit layout is generated in ARES. The simulation clock rate for the MCU should be set in the component properties dialog when the COF file is attached; 4 MHz is usually used in the demo circuits, giving an instruction cycle time of 1 μs. This determines the programmed delay count required to give the specified output rate. If the output LSB frequency is 50 Hz, the period is 20 ms. The half-cycle time then is 10 ms, which is the required program delay.

Appendix B explains the program design process in more detail.

Software Design Using CCS C

A program is to be designed to meet the specification given in Appendix A, which describes how to develop the hardware design for this application. The specification was as follows: When a button is pressed, the system generates an 8-bit binary count, starting at 0, on a bar graph display. The output frequency at the least significant bit is 50 Hz, giving an overall cycle time of 2.56 sec.

The general form of a real-time application is represented by the flowchart in Figure B.1, which shows two main phases: initialization and main loop. The initialization is executed once, and the main loop repeats.

The program must be written to the syntax requirements of standard C, with reference in this case to the CCS C User Manual (Version 4), downloadable from as a PDF from www.ccsinfo.com. The dialect of C developed by CCS Inc. is tailored specifically to the features of the PIC microcontroller. CCS supplies different complier variants for low-, middle-, and high-performance PICs; the mid-range compiler PCM is used for the PIC 16F877A.

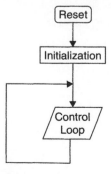

Figure B.1: Real-Time Application Flowchart

The initialization phase typically contains statements that include the MCU-specific header and library files specific to the target device. The main program is contained in a function `main()`. Variables and data structures defined at this point are global in scope (recognized and unique throughout the whole program). The endless loop can be started with `while(1)` or `for(;;)`, both of which mean to run an endless loop (unconditionally).

The main loop contains various conditional sequences and loops, comprising data operations and function calls. These functions may be built into the compiler, included as additional libraries with the `use` directive, or written by the user. They process input or stored data and return results to be used by later functions, for example, as system output. A general outline of a C program is shown in Listing B.1.

Listing B.1   C Program General Outline

```
Header comment block
Include resource files
Other preprocessor commands

Function blocks
   Function name(plus received parameters)
   Local variable & data structure declarations
   Unconditional sequences
   Conditional sequences
   Loop sequences
   Function calls
   Return to calling block with results

Main block
   Variable declarations
   Data structure definitions
   Loop
      Unconditional Sequences
      Conditional Sequences
      Loop Sequences
      Function Calls
Endlessly
```

BAR1 Source Code

The program source code (Listing B.2) starts with a comment block containing the name of the project, author, date, version, and program description. Details of the compiler version, development system, and target hardware can be included. In other words, as much information as possible to allow the code to be modified, updated, and maintained effectively. In CCS C, the initialization phase includes a header file that defines the MCU for which the program is intended. This is necessary as every PIC processor has

---

Listing B.2   Source Code BAR1.C

```
/* HEADER COMMENT SECTION ****************************************

  BAR1.C MPB V1.0  Source code file details
  Output binary count  Program description
  Simulation version  Target system details */

// INITIALIZATION SECTION ****************************************

#include "16F877A.h"                  // Define MCU regsisters etc
#use delay (clock = 4000000)          // Include delay routines

void main()                           // Define main program block
{                                     // Start of main block
  int x;                              // Declare variable

// CONTROL LOOP SECTION ****************************************

  while(1)                            // Define endless loop
  {                                   // Start of main loop
    if(!input(PIN_A4))                // Test input button
    {                                 // Start of conditional block
        output_C(x);                  // Output binary code
        x++;                          // Increment output variable

    }                                 // End of conditional block
    delay_ms(10);                     // Wait 10ms
  }                                   // End of main loop
}                                     // End of program
```

---

a different set of features: the number of ports, memory size, special input and output facilities, and so on. The include statement is defined as a compiler (preprocessor) directive by the leading hash symbol (#). The include directive inserts the source code from the specified file as though it had been typed in. Your own files can be included, so you can make a library of your own routines for reuse as required.

Many built-in functions are included by the compiler automatically, for example, output_C(x). Others have to be specified with use, which identifies a library of functions used later in the program. The directive #uses delay (clock=4000000) calls up the set of delay routines that need the MCU clock speed to be stated so that the correct delays can calculated. The compiler manual indicates which functions need to be preceded by a use directive.

The initialization phase includes defining all global variables. The variable labels, such as x or input_value, are attached to the address where the variable value is to be stored. The variable type declaration (e.g., int) allows the compiler to allocate an appropriate set of locations for the variable. In CCS C, the default integer size is 8 bits, in others it is 16. Global variables remain in existence while the program is running and are recognized throughout all levels of the program.

However, to save data memory and allow some duplication of labels, local variables may be defined within a function. These then exist only for the duration of the function execution and are subsequently lost. The value of local variables can be passed back to the calling function or should be defined as global, so that the data are not lost when the function completes.

PIC Registers

Some knowledge of the PIC internal architecture is useful at this point. The MCU operation is controlled by a set of file registers, which contain special function registers (SFRs) in the first 32 locations, followed by some general purpose registers (GPRs). The 16F877 has four banks of 128 registers, as shown in Figure B.2. Some registers are duplicated in more than one bank, so the actual number of distinct GPRs is 192.

Figure B.3 shows the function of each bit of the SFRs in Bank0 and Figure B.4 the details for the status register, which contains the bank select bits. Note that the file register

Register	File Address	Register	File Address	Register	File Address	Register	File Address
Indirect addr. (*)	00h	Indirect addr. (*)	80h	Indirect addr. (*)	100h	Indirect addr. (*)	180h
TMR0	01h	OPTION_REG	81h	TMR0	101h	OPTION_REG	181h
PCL	02h	PCL	82h	PCL	102h	PCL	182h
STATUS	03h	STATUS	83h	STATUS	103h	STATUS	183h
FSR	04h	FSR	84h	FSR	104h	FSR	184h
PORTA	05h	TRISA	85h		105h		185h
PORTB	06h	TRISB	86h	PORTB	106h	TRISB	186h
PORTC	07h	TRISC	87h		107h		187h
PORTD(1)	08h	TRISD(1)	88h		108h		188h
PORTE(1)	09h	TRISE(1)	89h		109h		189h
PCLATH	0Ah	PCLATH	8Ah	PCLATH	10Ah	PCLATH	18Ah
INTCON	0Bh	INTCON	8Bh	INTCON	10Bh	INTCON	18Bh
PIR1	0Ch	PIE1	8Ch	EEDATA	10Ch	EECON1	18Ch
PIR2	0Dh	PIE2	8Dh	EEADR	10Dh	EECON2	18Dh
TMR1L	0Eh	PCON	8Eh	EEDATH	10Eh	Reserved(2)	18Eh
TMR1H	0Fh		8Fh	EEADRH	10Fh	Reserved(2)	18Fh
T1CON	10h		90h		110h		190h
TMR2	11h	SSPCON2	91h		111h		191h
T2CON	12h	PR2	92h		112h		192h
SSPBUF	13h	SSPADD	93h		113h		193h
SSPCON	14h	SSPSTAT	94h		114h		194h
CCPR1L	15h		95h		115h		195h
CCPR1H	16h		96h	General	116h	General	196h
CCP1CON	17h		97h	Purpose	117h	Purpose	197h
RCSTA	18h	TXSTA	98h	Register	118h	Register	198h
TXREG	19h	SPBRG	99h	16 Bytes	119h	16 Bytes	199h
RCREG	1Ah		9Ah		11Ah		19Ah
CCPR2L	1Bh		9Bh		11Bh		19Bh
CCPR2H	1Ch	CMCON	9Ch		11Ch		19Ch
CCP2CON	1Dh	CVRCON	9Dh		11Dh		19Dh
ADRESH	1Eh	ADRESL	9Eh		11Eh		19Eh
ADCON0	1Fh	ADCON1	9Fh		11Fh		19Fh
General Purpose Register 96 Bytes	20h ... 7Fh	General Purpose Register 80 Bytes	A0h ... EFh	General Purpose Register 80 Bytes	120h ... 16Fh	General Purpose Register 80 Bytes	1A0h ... 1EFh
		accesses 70h–7Fh	F0h ... FFh	accesses 70h–7Fh	170h ... 17Fh	accesses 70h–7Fh	1F0h ... 1FFh
Bank 0		Bank 1		Bank 2		Bank 3	

Figure B.2: PIC 16F877 File Registers (by permission of Microchip Technology Inc.)

Address	Name	Bit 7	Bit 6	Bit 5	Bit 4	Bit 3	Bit 2	Bit 1	Bit 0	Value on: POR, BOR	Details on page:
Bank 0											
00h[3]	INDF	Addressing this location uses contents of FSR to address data memory (not a physical register)								0000 0000	31, 150
01h	TMR0	Timer0 Module Register								xxxx xxxx	55, 150
02h[3]	PCL	Program Counter (PC) Least Significant Byte								0000 0000	30, 150
03h[3]	STATUS	IRP	RP1	RP0	$\overline{TO}$	$\overline{PD}$	Z	DC	C	0001 1xxx	22, 150
04h[3]	FSR	Indirect Data memory Address Pointer								xxxx xxxx	31, 150
05h	PORTA	—	—	PORTA Data Latch when written: PORTA pins when read						--0x 0000	43, 150
06h	PORTB	PORTB Data Latch when written: PORTB pins when read								xxxx xxxx	45, 150
07h	PORTC	PORTC Data Latch when written: PORTC pins when read								xxxx xxxx	47, 150
08h[4]	PORTD	PORTD Data Latch when written: PORTD pins when read								xxxx xxxx	48, 150
09h[4]	PORTE	—	—	—	—	—	RE2	RE1	RE0	---- -xxx	49, 150
0Ah[1,3]	PCLATH	—	—	—	Write Buffer for the upper 5 bits of the Program Counter					---0 0000	30, 150
0Bh[3]	INTCON	GIE	PEIE	TMR0IE	INTE	RBIE	TMR0IF	INTF	RBIF	0000 000x	24, 150
0Ch	PIR1	PSPIF[3]	ADIF	RCIF	TXIF	SSPIF	CCP1IF	TMR2IF	TMR1IF	0000 0000	26, 150
0Dh	PIR2	—	CMIF	—	EEIF	BCLIF	—	—	CCP2IF	-0-0 0--0	28, 150
0Eh	TMR1L	Holding Register for the Least Significant Byte of the 16-bit TMR1 Register								xxxx xxxx	60, 150
0Fh	TMR1H	Holding Register for the Most Significant Byte of the 16-bit TMR1 Register								xxxx xxxx	60, 150
10h	T1CON	—	—	T1CKPS1	T1CKPS0	T1OSCEN	$\overline{T1SYNC}$	TMR1CS	TMR1ON	--00 0000	57, 150
11h	TMR2	Timer2 Module Register								0000 0000	62, 150
12h	T2CON	—	TOUTPS3	TOUTPS2	TOUTPS1	TOUTPS0	TMR2ON	T2CKPS1	T2CKPS0	-000 0000	61, 150
13h	SSPBUF	Synchronous Serial Port Receive Buffer/Transmit Register								xxxx xxxx	79, 150
14h	SSPCON	WCOL	SSPOV	SSPEN	CKP	SSPM3	SSPM2	SSPM1	SSPM0	0000 0000	82, 82, 150
15h	CCPR1L	Capture/Compare/PWM Register 1 (LSB)								xxxx xxxx	63, 150
16h	CCPR1H	Capture/Compare/PWM Register 1 (MSB)								xxxx xxxx	63, 150
17h	CCP1CON	—	—	CCP1X	CCP1Y	CCP1M3	CCP1M2	CCP1M1	CCP1M0	--00 0000	64, 150
18h	RCSTA	SPEN	RX9	SREN	CREN	ADDEN	FERR	OERR	RX9D	0000 000x	112, 150
19h	TXREG	USART Transmit Data Register								0000 0000	118, 150
1Ah	RCREG	USART Receive Data Register								0000 0000	118, 150
1Bh	CCPR2L	Capture/Compare/PWM Register 2 (LSB)								xxxx xxxx	63, 150
1Ch	CCPR2H	Capture/Compare/PWM Register 2 (MSB)								xxxx xxxx	63, 150
1Dh	CCP2CON	—	—	CCP2X	CCP2Y	CCP2M3	CCP2M2	CCP2M1	CCP2M0	--00 0000	64, 150
1Eh	ADRESH	A/D Result Register High Byte								xxxx xxxx	133, 150
1Fh	ADCON0	ADCS1	ADCS0	CHS2	CHS1	CHS0	$\overline{GO/DONE}$	—	ADON	0000 00-0	127, 150

Legend: x = unknown, u = unchanged, q = value depends on condition, – = unimplemented, read as '0', r = reserved. Shaded locations are unimplemented, read as '0'.

Figure B.3: PIC 16F877 Registers, Bank 0 (by permission of Microchip Technology Inc.)

bank select bits RP0 and RP1 are used for direct addressing, but IRP is used for indirect addressing via the file select register (FSR).

In this case, the value in the register specified in the FSR is read or written at file address 00. The PIC internal architecture and register operations are fully explained in the 16F87XA data sheet downloadable from www.microchip.com.

R/W-0	R/W-0	R/W-0	R-1	R-1	R/W-X	R/W-X	R/W-X
IRP	RP1	RP0	$\overline{\text{TO}}$	$\overline{\text{PD}}$	Z	DC	C

bit 7   bit 0

bit 7   **IRP**: Register Bank Select bit (used for indirect addressing)
1 = Bank 2, 3 (100 h-1 FFh)
0 = Bank 0, 1 (00 h-FFh)

bit 6–5   **RP1:RP0**: Register Bank Select bits (used for direct addressing)
11 = Bank 3 (180 h-1FFh)
10 = Bank 2 (100 h-17Fh)
01 = Bank 1 (80 h-FFh)
00 = Bank 0 (00 h-7 Fh)
Each bank is 128 bytes.

bit 4   **$\overline{\text{TO}}$**: Time-out bit
1 = After power-up, CLRWDT instruction or SLEEP instruction
0 = A WDT time-out occurred

bit 3   **$\overline{\text{PD}}$**: Power-down bit
1 = After power-up or by the CLRWDT instruction
0 = By execution of the SLEEP instruction

bit 2   **Z**: Zero bit
1 = The result of an arithmetic or logic operation is zero
0 = The result of an arithmetic or logic operation is not zero

bit 1   **DC**: Digit carry/borrow bit (ADDWF, ADDLW, SUBLW, SUBWF instructions)
(for borrow, the polarity is reversed)
1 = A carry-out from the 4th low order bit of the result occurred
0 = No carry-out from the 4th low order bit of the result

bit 0   **C**: Carry/borrow bit (ADDWF, ADDLW, SUBLW, SUBWF instructions)
1 = A carry-out from the Most Significant bit of the result occurred
0 = No carry-out from the Most Significant bit of the result

         Note: For borrow, the polarity is reversed. A subtraction is executed by adding the two's complement of the second operand. For rotate (RRF, RLF) instructions, this bit is loaded with either the high, or low order bit of the source register.

Figure B.4: PIC 16F877 Status Register Bit Functions (by permission of Microchip Technology Inc.)

BAR1 List File

The list file BAR1.LST, in Listing B.3, shows the assembly language version of the program produced by the compiler. This book does not assume knowledge of assembler programming, but for those readers who have followed the usual progression from assembler, the list file gives a useful insight into how the compiler works. Comments (italics) have been added to the original file to explain its operation. The original source code is highlighted in bold.

The compiler initially sets the memory page to 0 by loading the PCLATH (program counter latch high) register (0 A) with 0. This is the reset default setting anyway, but

Listing B.3   List File BAR1.LST

```
CCS PCM C Compiler, Version 4.024, 37533 16-Feb-07 17:05

   Filename: bar1.lst

   ROM used: 59 words (1%)
            Largest free fragment is 2048
   RAM used: 8 (2%) at main() level
            9 (2%) worst case
   Stack:     1 locations

*; START OF INITIALISATION ****************************
0000: MOVLW 00
0001: MOVWF 0A    ; Select Program Page 0
0002: GOTO 01B    ; Jump to main block
0003: NOP
.................... /*    BAR1.C MPB V1.0
....................       Output binary count
....................       when button pressed
....................       LSB = 50Hz
....................       Simulation version
.................... */
....................
.................... #include "16F877A.h"
.................... //////// Standard Header file for the PIC16F877A
device ///////////////////
.................... #device PIC16F877A
.................... #list
....................
....................            ; FUNCTION ROUTINE ********************
....................
.................... #use delay (clock = 4000000)
0004: MOVLW 22
0005: MOVWF 04           ; Point to delay value
0006: BCF 03.7           ; Select File Bank 0,1 indirect addressing
0007: MOVF 00,W          ; Fetch delay value
0008: BTFSC 03.2         ; If delay value = 0...
0009: GOTO 018           ; ...skip this routine
000A: MOVLW 01           ; START 1ms DELAY LOOP
```

```
000B: MOVWF 78              ; Load delay value high byte = 01
000C: CLRF 77               ; Load delay value low byte = 00
000D: DECFSZ 77,F           ; Decrement low counter...
000E: GOTO 00D              ; ...and repeat x255 = 765us
000F: DECFSZ 78,F           ; Decrement high counter...
0010: GOTO 00C              ; ...and do not repeat
0011: MOVLW 4A              ; Load low counter...
0012: MOVWF 77              ; ...with 0x4A(74)
0013: DECFSZ 77,F           ; Decrement low counter...
0014: GOTO 013              ; ...and repeat x73 = 219us
0015: GOTO 016              ; Next step
0016: DECFSZ 00,F           ; Decrement delay value...
0017: GOTO 00A              ; ...and repeat 1ms delay loop x9
0018: BCF 0A.3              ; Select program memory page zero
0019: BCF 0A.4
001A: GOTO 039 (RETURN)     ; Jump back to main block
...................
...................
...................              ; START OF MAIN BLOCK **************
................... void main()
................... {
001B: CLRF 04               ; Set FSR pointer = 0
001C: BCF 03.7              ; Select File Bank 0,1 for indirect
                              addressing
001D: MOVLW 1F
001E: ANDWF 03,F            ; Select File Bank 0 for direct
                              addressing
001F: BSF 03.5              ; Select File Bank 1
0020: BSF 1F.0              ; Select analogue input mode 8
0021: BSF 1F.1
0022: BSF 1F.2
0023: BCF 1F.3
0024: MOVLW 07              ; Switch off comparator inputs
0025: MOVWF 1C
................... int x;   ; File register 0x21 (GPR1)
                              appointed as x
...................
...................
...................              ; START OF MAIN LOOP
...................
```

```
.................... while(1)              ; place GOTO 0x29 at main loop end
.................... {
.................... if(!input(PIN_A4))
*
0029: BSF 03.5                             ; Select file bank 1
002A: BSF 05.4                             ; Set RA4 as input
002B: BCF 03.5                             ; Select file bank 0
002C: BTFSC 05.4                           ; Test input RA4...
002D: GOTO 036                             ; and skip next block if high
.................... {
.................... output_C(x);
*
0026: MOVLW FF
0027: BCF 03.5                             ; Select file bank 0
0028: MOVWF 20                             ; GPR0 = 0xFF
*
002E: MOVLW 00                             ; GPR0 = 0 x 00
002F: MOVWF 20
0030: BSF 03.5                             ; Select file bank 1
0031: CLRF 07                              ; Port C = output
0032: BCF 03.5                             ; Select file bank 0
0033: MOVF 21,W                            ; Output x
0034: MOVWF 07
.................... x++;
0035: INCF 21,F                            ; Increment x
.................... }
.................... delay_ms(10);
0036: MOVLW 0A                             ; Load delay value...
0037: MOVWF 22                             ; into GPR2
0038: GOTO 004                             ; Jump to delay routine
.................... }
0039: GOTO 029                             ; Jump back to start of main loop
....................
.................... }
003A: SLEEP                                ; Shut down (not normally executed)
Configuration Fuses:
  Word 1: 3F73 RC NOWDT PUT NODEBUG NOPROTECT BROWNOUT NOLVP NOCPD NOWRT
```

the compiler does not rely on this. The format of the file registers in Bank 0 is shown in Figure B.2. The program then jumps over the delay function block.

The main block starts by initializing the memory bank selection and the analog inputs. The variable x is then declared and the compiler allocates file register 0x21 (GPR1) as its storage location. The statement while(1) at the start of the main loop instructs the compiler to place a GOTO at the end of the loop with the address of the first loop instruction as its destination (address 0x29).

The if() statement is implemented by first setting the pin RA4 as input then testing it. We can see here that the pin initialization is repeated every time the statement is executed. This is an example of an operation where C is clearly less efficient than assembler, where the pin would normally be initialized once only. The same problem occurs in the next block, when the value of x is output—the initialization is repeated each time the statement is executed.

The delay period (10) is stored in the next available location, 0x22, when the delay is called. The program then jumps back to the delay code block starting address 0x04. A counting sequence follows, which gives a delay of 1 ms. This is repeated ten times, and the program jumps back to the main block and the main loop repeats. Note that assembler instructions CALL and RETURN are not used, because this would limit the number of nested routines to eight, the limit of the PIC stack depth. By using GOTO instead, this limitation is avoided by the CCS complier.

System Testing Using Proteus VSM

A hardware design schematic BAR1.DSN has been devised (Appendix A) and an application program BAR1.C developed (Appendix B) from the specification. These can now be brought together for testing in simulation mode.

Attaching the Program

The application program is output by the C compiler as a file called BAR1.COF, which should be stored in the project directory BAR1. It contains the machine code plus some debugging information required by the simulator to display the program source code. Several other files are created by the compiler at the same time, and all these should be stored in the same project folder containing the ISIS design file BAR1.DSN.

On the schematic, right click, then left click on the PIC chip to display the component properties (Figure C.1). The folder browse button allows the COF file to be opened (attached) to the virtual processor, and the MCU clock frequency can be set at the same time. The 4 MHz is a useful default clock frequency, as this gives a 1-μs instruction cycle time and is the maximum frequency using a standard crystal (XT mode in the chip configuration settings). This clock setting must be passed to the delay routine in the program. The Program Configuration Word has no significant effect at this stage but must be set as appropriate when programming real hardware.

Program Debugging

The program can be run by pressing the Play button in the control console. If the program is correct, the specified output is seen. The bar graph displays a binary count when

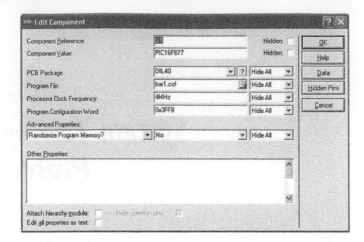

Figure C.1: MCU Properties Dialog for Attaching the Program

the Input button on the schematic is "pressed" using the mouse pointer. It should take 2.56 sec to cycle through all the output codes with a loop delay of 10 ms. This can be checked using the simulation clock at the bottom of the screen.

If the program does not work as required, it needs debugging. The screenshot in Figure C.2 shows some of the debugging features. The principal technique is single stepping—the program sequence is checked by executing one statement at a time. This requires the source code to be displayed; pause the program and select the Debug menu, PIC CPU Source Code. The source code window appears, with the current execution point highlighted. If the Pause control is pressed instead of Run, the program can be single stepped from the first statement. This is useful if the initialization sequence needs to be checked.

It is not possible to operate the debugging tools and the interactive push button with the mouse at the same time. Therefore, in Figure C.2, the Input button is shorted out with a temporary link so that the output runs continuously. Alternatively, it can be replaced with a switch for simulation purposes.

The source code window has a selection of debug buttons: Run, Step Over, Step Into, Step Out Of . . . the current function. Step Over means execute the following function call at full speed, stopping on return, while Step Into means execute the function stepwise. While stepping through a function, Step Out Of allows you to return to the calling block at full speed. This is useful for getting out of a function you have inadvertently stepped into.

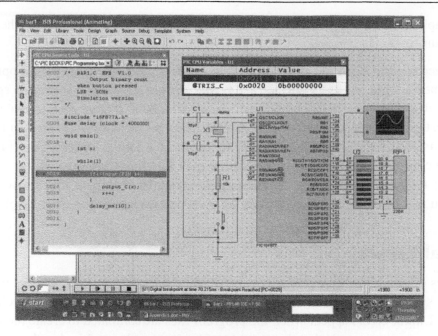

Figure C.2: Program Simulation Screenshot

The Breakpoint button is used to set and clear breakpoints in the code at the current cursor position. Program execution is run at full speed, until stopped at the breakpoint. Additional breakpoint control options can be selected by right clicking on the source code window. This source window menu also allows the display to be modified to show line numbers and program memory addresses. The assembler code for each statement can be displayed by selecting Disassembly. Note that several lines of assembler code are needed for each C statement—this is the reason that the C program needs more memory. The Set Font option is useful if displaying the PC screen on a projector (teachers note); the text can be enlarged for better visibility.

PIC CPU variables are displayed from the Debug menu. Right click on the window and deselect the Globals option, leaving just the program variables visible. The display numerical format can then be changed by right clicking on the variable in the window, for example, to display the variables as unsigned integers if only positive whole numbers are used.

The CPU registers may be displayed if required, as well as the CPU data memory, that is, the file registers. Some of these have special or system functions, the rest are available for variable storage. Remember that some variable types use more than one location;

for example, a 16-bit integer uses two. The variable locations are highlighted when they change during single stepping.

If you need to slow down the program execution, go to the System, Animation options. The Frames per Second and Timestep per Frame settings control the simulation speed. The default settings are 20 f/s and 50 ms/f, giving $20 \times 50 = 1000$ ms/sec, or real time. If the Timestep per Frame is reduced to, say, 5 ms, the simulation slows down by a factor of 10. This allows the system operation to be observed at a more leisurely pace in Run mode. In complex applications, the simulation may slow down automatically to allow the processor to complete the circuit solution for each simulation step, in which case, it does not run in real time. This can be checked by observing the simulation clock display.

Typical Errors

The types of errors that appear when the program is compiled are either syntax or linker errors. A syntax error might be a spelling mistake in the source code or an undeclared variable. Linker errors appear when the program files are combined to create the final program; a common one is that the `include` files have not been placed in the project folder and cannot be found by the linker.

Logical errors, on the other hand, appear only when the program is tested; and these are easier to correct if detected prior to downloading to hardware, by using a simulator such as MPSIM or VSM. VSM is easier to use, as the errors are more readily spotted in the animated schematic than in the numerical output of MPSIM.

Some simple examples of possible errors in BIN1.C are outlined next.

Sequence Error

While the increment statement follows the output statement, the first output is 00000000. If, instead, the increment were placed before the output, the first output is 00000001, and this is not as specified. This error is not evident in the Run mode but is detected if the program is single stepped from the top (hit Pause initially rather than Run).

Inversion Error

This is a logical error that causes the opposite effect to that required. For example, if the exclamation mark is omitted before the input function, the output runs when the button is open rather than closed.

Parameter Error

If the wrong input is specified in the input statement (e.g., PIN_A5 instead of PIN_A4), the button has no effect, as the wrong input is being tested. This error is detected by comparing the program and schematic.

Timing Error

The delay time is calculated so that the LSB toggles every 10 ms. If this figure is incorrect, the output frequencies are wrong. This can be checked by using the simulation clock or a virtual oscilloscope.

The simulation clock is displayed at the bottom of the schematic window. To check the period of the output, a breakpoint can be set at the beginning of the main loop. The program then stops once per cycle, and the time taken per cycle can be read from the clock. A breakpoint is set by clicking on the Breakpoint button at the top of the source code widow.

The oscilloscope allows the output to be displayed in the time domain. It is selected from the Virtual Instruments list. Input A should be connected to the output, RC0, and

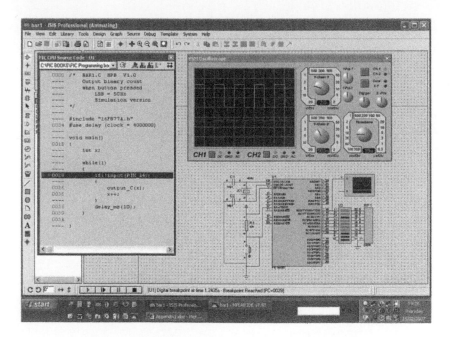

Figure C.3: Virtual Oscilloscope Screenshot

a full-size version of the scope should appear when the simulation is run. If not, enable it in the Debug menu. Adjust the controls to see the 50-Hz waveform displayed.

Figure C.3 shows the VSM analog scope and simulation clock display. A breakpoint has been set at the `if` statement, so the clock increments by 10 ms each time Run is selected. ISIS also provides virtual signal sources, meters, voltage and current probes, logic analyzer, and counter/timer, as well as a graphing feature for analog and digital signals. When the program is fully debugged, it can be downloaded to hardware and retested. This should leave only hardware faults to be rectified to obtain a working system.

Readers should note that Proteus VSM is continuously updated. New features and components are added on a regular basis. Specifically, new MCUs are added as they are released by the manufacturers. Version 6 was used to produce the simulation circuits in this book. Version 7 has since been released, which has, for example, an enhanced 4-channel virtual oscilloscope. Visit www.labcenter.co.uk for the latest product information.

C Compiler Comparison

The intention of this book is to introduce C programming for all microcontrollers. However, particular products have to be selected to act as examples. When the basics have been explained using one particular combination of MCU, compiler, and development system, others can be considered.

The CCS C compiler was selected for this book principally because it has an extensive library of peripheral driver routines, is reasonably inexpensive, and is recognized by Microchip and Labcenter as a preferred compiler. However, several other suitable compilers are available at the time of this writing, so it would be useful to see how they compare. The following products have been selected, but bear in mind that, in the rapidly moving microcontroller market, significant changes probably have occurred by the time you read this:

- Microchip C18
- HiTech PIC C
- Mikroelektronika C
- Matrix Multimedia C

The first two are professional compilers, which would tend to be used by more experienced engineers. The second two are aimed at the educational market and include more user-friendly features to help the beginner.

Other PIC C compilers are available that are not considered here. They are typically supplied by companies that produce development tools for a range of different processors, which could suit application developers who use a range of MCU types. They do not provide the range of library functions considered essential here.

Each compiler has a set of header files provided, all of which have a similar function of defining the register and control bit labels for all the supported processors. The exact labeling system can vary, although the labeling used in the PIC hardware manuals must be preferred.

Microchip C18

Microchip does not supply a compiler for the mid-range 16 series MCUs. It is assumed that any application developed in C will be run on an 18 series processor or above. This is because the mid-range devices have limited memory capacity, and many commercial C applications exceed this limit.

Nevertheless, it is well worth looking at C18, because having learned C on the 16 series, the reader may wish to consider the option of progressing to the 18 series for further work. The full list of features claimed for this compiler, as listed in the *C18 User Guide* (www.microchip.com) includes

- ANSI '89 compatibility.

- Integration with the MPLAB IDE for easy-to-use project management and source-level debugging.

- Generation of relocatable object modules for enhanced code reuse.

- Compatibility with object modules generated by the MPASM assembler, allowing complete freedom in mixing assembly and C programming in a single project.

- Transparent read/write access to external memory.

- Strong support for in-line assembly when total control is absolutely necessary.

- Efficient code generator engine with multilevel optimization.

- Extensive library support, including PWM, SPI™, I²C™, UART, USART, string manipulation, and math libraries.

- Full user-level control over data and code memory allocation.

It must be assumed that the integration of C18 into the MPLAB IDE will be reasonably seamless, giving it a built-in advantage over competing compilers. Source-level debugging, in particular, can reveal limitations in the effectiveness of the integration into the IDE of a third party product.

Relocatable object modules allow the user to build up a library of reusable routines. This is obviously useful when producing a series of similar application programs. If particular hardware peripherals are used repeatedly in different designs, the same driver routines, perhaps with minor variations, can be used. However, these routines must be designed to receive and return variable values in a consistent manner to maximize the benefits of this approach.

Library routines are provided for the main peripheral interfaces, and a comprehensive selection is found in the *C18 Compiler Libraries* manual. Software drivers allow peripherals to be connected to any pin, not just those associated with the internal hardware interface. This provides more flexibility in the use of the chip pins and may mean that a cheaper device can be used for a particular application.

If we look at some source code examples provided in the *C18 User Guide*, we may be able to identify some of the features where C18 and CCS C diverge. Remember, however, that the general language syntax must conform to the ANSI standard. Listing D.1 shows a simple LED flasher program.

Listing D.1   C18 Sample Source Code (LED Flasher)

```
#include <p18cxxx.h>              /* MCU header file **********/

  void delay (void)               /* Delay function ************/
  {
      unsigned int i;
      for (i=0; i<10000; i++);
  }


void main (void)                  /* Main Program **************/
{
  TRISB = 0;                      /* Port B output                  */
  while(1)                        /* Loop always                    */
  {
      PORTB = 0;                  /* Reset the LEDs                 */
      delay();                    /* Delay to see change            */
      PORTB = 0x5A;               /* Light the LEDs                 */
      delay();                    /* Delay to see change            */
  }
}
```

The MCU header file is included in the same way as in CCS C, and the delay routine uses standard syntax. The main difference evident is that the port registers are addressed directly by assigning a value to the data direction register (e.g., TRISB=0) and the output data register (e.g., PORTB=0x5A). In CCS C, a function is used (output_B(0)). The C18 syntax is arguably simpler.

Listing D.2, a C18 program using interrupts, illustrates some other differences. As in many PIC C compilers, direct access to the register control bits is used, for example, in the statement INTCONbits.TMR0IF=0, which resets the timer interrupt flag. This requires knowledge of the internal architecture, which makes the programming more difficult. CCS C sensibly avoids the need for such direct access. The timer setup statement uses a function call in a similar format to CCS, but of course, the exact syntax is different.

Listing D.2 also includes other features not covered elsewhere in this book. The #pragma directive allows additional directives to be defined for this specific compiler and added to the standard set defined in the ANSI standard. The keywords _asm and _endasm enclose a section of assembly language code, in this case just one instruction GOTO label.

Hi-Tech PIC C

The Hi-Tech PIC C is a professional standard compiler supplied by a company well established as a development system tool supplier. Hi-Tech supplies C compilers for wide range of microcontrollers on the market: PIC 16, 18, 24, and dsPIC (digital signal processors) as well as Freescale 68000-based types, ARM, 8051 derivatives, Texas Instruments MSP430 devices, and other legacy products.

The features claimed are these:

- ANSI C—full featured and portable.

- Reliable—mature, field-proven technology.

- Multiple C optimization levels.

- An optimizing assembler.

- Full linker, with overlaying of local variables to minimize RAM usage.

- Comprehensive C library with all source code provided.

Listing D.2 C18 Sample Source Code (LED Output Using Timer Interrupt)

```
#include <p18cxxx.h>
#include <timers.h>

#define NUMBER_OF_LEDS 8

void timer_isr (void);
static unsigned char s_count = 0;

#pragma code low_vector=0x18

  void low_interrupt (void)
  {
      _asm GOTO timer_isr _endasm
  }

#pragma code
#pragma interruptlow timer_isr

  void timer_isr (void)
  {
      static unsigned char led_display = 0;
      INTCONbits.TMR0IF = 0;
      s_count = s_count % (NUMBER_OF_LEDS + 1);
      led_display = (1 << s_count++) - 1;
      PORTB = led_display;
  }

void main (void)
{
  TRISB = 0;
  PORTB = 0;

  OpenTimer0 (TIMER_INT_ON & T0_SOURCE_INT & T0_16BIT);
  INTCONbits.GIE = 1;

  while (1) {}
}
```

- Support for 24-bit and 32-bit IEEE floating point and 32-bit long data types included.

- Mixed C and assembler programming.

- Unlimited number of source files.

- Listings showing generated assembler.

- Compatible—integrates into the MPLAB® IDE, MPLAB ICD, and most third party development tools.

- Runs on multiple platforms: Windows®, Linux®, UNIX®, Mac OS X, Solaris™.

Optimization involves reducing the final code size by removing redundant code and modifying the assembler version to reduce the number of instructions to the minimum achievable.

The most obvious disadvantage of this compiler is that only the standard library functions for data conversion, memory management, mathematical operations, and basic I/O are provided. It is assumed that the user will develop the peripheral drivers as required, to suit the particular range of applications and hardware to be supported, or that the peripheral control registers will be accessed directly.

On the other hand, a major advantage is that a fully featured freeware version, PICC-Lite, is available for hobbyists, students, and limited commercial purposes. At the time of writing, the following PIC MCUs are supported with no limitations, as compared to the full version: 12F629, 12F675, and 16F84. A further set of 16 series chips can be used with a limitation on RAM and program memory: '627, '684, '690, '877, '887, and '917. Other limitations are imposed due to the limited memory available in these chips.

Hi-Tech also supplies Salvo RTOS, including a freeware version. This is a cooperative, event-driven, priority-based, multitasking, real-time operating system designed for microcontrollers with limited RAM and ROM. The manual supplied (www.pumpkininc. com) with this product contains a very useful introduction to RTOS principles and is recommended if further information is required on using RTOS in PICs.

An example of Hi-Tech C source code is shown in Listing D.3. It outputs a binary count at Port B that is incremented every second using a timer interrupt. The port register is addressed directly, using the label PORTB. The timer control bit labels are defined in the header file PIC.H and set directly in the main routine. Note that here the calculation of the initial loop count constant RELOADS is calculated in the initial directive block using the arithmetic and logic operations provided within the directive syntax. Recall that CCS C uses a directive to declare a function as an ISR; here, the compiler recognizes the keyword interrupt within the function name instead.

Listing D.3   Hi-Tech C Sample Source Code (Timer Interrupt)

```c
#include <pic.h>

/*  Example code for using timer0 on a 16F84
    Sets up a 1 second interrupt and increments Port B
*/

/* Calculate preload value for one second timer ************/
#define  PERIOD  1000000              // Period in us-one second here
#define  XTAL  4000000                // Crystal frequency-4MHz
#define  IPERIOD  (4 * 1000000/XTAL)  // Period of instruction clock in us
#define  SCALE  256                   // Timer 0 prescaler
#define  T0_TICKS 256                 // Number of counts for interrupt
#define  TICK_PERIOD (SCALE * IPERIOD) // Period (us) of timer clock
#define  RELOADS((PERIOD/T0_TICKS)/   // Calculate preload value
TICK_PERIOD)

unsigned long seconds;               // Second count
near char    reload = 0;             // Reload count

/* Service routine for timer 0 interrupt ******************/
void interrupt timer0_isr(void)      // Define function as timer ISR
{
  if(reload == 0){
    reload = RELOADS + 1;            // Set initial value of reload
                                     // count
    seconds++;                       // Count seconds
    PORTB++;                         // Change port display
  }
  reload--;                          // Count down reloads
  T0IF = 0;                          // Clear timer interrupt flag
}
main()                               /* Initialise timer and wait for
                                        interrupt ************/
{
  OPTION = 0b0111;                   // prescale by 256
  T0CS = 0;                          // select internal clock
  T0IE = 1;                          // enable timer interrupt
  GIE = 1;                           // enable global interrupts
  TRISB = 0;                         // output changes on LED

  for(;;)
  continue;                          // let interrupt do its job
}
```

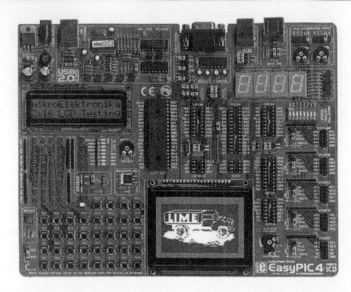

Figure D.1: Mikroelectronica EasyPIC4 Development Board

Mikro C

Mikroelectronica supplies range evaluation and development boards for the PIC and other microcontrollers, as well as C, Pascal, and Basic compilers (Figure D.1). The C compiler MikroC is well documented in a downloadable user manual and includes a good range of peripheral driver libraries, including CAN, Ethernet, and graphical LCD drivers as part of a comprehensive I/O library. The packages are oriented toward the educational and hobby market, offering additional features designed to assist the beginner in developing C applications.

An evaluation version does not appear to be available at the time of this writing, and the compiler syntax can be assessed prior to purchase only by reference to code fragments given in the manual. An ADC input block is reproduced as an example in Listing D.4. As we see, the control registers are set up by loading control codes as hex numbers, which requires the program designer to look up the necessary bit configurations. However, the ADC access function is simple and concise, allowing the input channel to be selected as the function parameter.

Matrix C

The primary product line of Matrix Multimedia is a user-friendly hardware system, E-blocks, that allows different systems to be assembled using plug-in modules. The

Listing D.4   MikroC Source Code Sample (ADC Input and Display)

```
unsigned inval;                      // 16-bit integer for 10-bit input

void main {
  ADCON1 = 0x80;                     // Setup ADC
  TRISA = 0xFF;                      // Analog inputs
  TRISB = 0x3F;                      // RB6,RB7 display outputs
  TRISD = 0;                         // Port D display outputs

  do{
    inval = Read_ADC(2);             // Read channel 2 (RA2)
    PORTD = inval;                   // Show low 8 bits
    PORTB = inval>>2;                // Show high 2 bits
  }while(1);
}
```

Figure D.2: Matrix Multimedia Modular PIC System

processor module incorporates sockets for a range of PIC MCUs and a number of D-type connectors. Peripheral modules with push buttons, LEDs, displays, keypad, relays, communications interfaces, and so on are added as required (Figure D.2).

Listing D.5   Matrix C Source Code Sample (ADC Input and Display)

```
#include <system.h>

void setupADC(void)
   {  trisb = 0x00;                /* Port B display               */
      trisa = 0xf1;                /* RA0 input, RA1-3 output      */
      adcon0 = 0x00;               /* Set up ADC                   */
      adcon1 = 0x80;               /* Set up ADC                   */
      ansel = 0x01;                /* Select AN0 only              */
   }

void main(void)
{
   setupADC();                     /* Call setup function          */

   while (1)                       /* Loop always                  */
   {   adcon0 = 0x05;              /* Start ADC                    */
       while(adcon0&0x04);         /* Wait until done              */
       portb = adresl;             /* Display low byte             */
       porta = adresh*2;           /* Display high bits            */
   }
}
```

The application programming can be implemented using a choice of assembler or C. Matrix also offers a proprietary flowchart-based programming system, Flowcode. The program is constructed using flowchart blocks, which are automatically converted to C and hence to assembler and machine code.

The C syntax used is illustrated in Listing D.5—a simple program to read an analog input and display the result. As in many C compilers for PIC, the control registers are loaded directly, and no special functions are used for peripheral access. The programming system is described via a tutorial, which is included with the compiler, so no separate reference manual is provided.

Summary of C Compilers

The features of the C compilers for the PIC 16 series MCUs outlined in this appendix are compared in Table D.1. We are particularly interested in using the 16F877, our reference device, which is used in the demo applications in the main part of this book. The compilers have been divided into commercial and educational categories.

Table D.1: Comparison of C Compilers for PIC 16 Series M

URL	Microchip C18 (microchip.com)	Hi-Tech C (htsoft.com)	CCS C (ccsinfo.com)	Mikro C (mikroe.com)	MM C (matrixmultimedia.com)
Primary market	Commercial	Commercial	Both	Educational	Educational
MCU targets	PIC 18 only	PIC + others	PIC only	Mainly PIC	Mainly PIC
Primary target hardware	Any	Any	Any	Proprietary single board	Proprietary modular
Function libraries	Extensive peripheral support	Standard libraries only	Good peripheral support	Extensive peripheral support	Standard libraries only
Tutorial or user manual	Comprehensive free download	Comprehensive free download	Free download	Comprehensive free download	Tutorial in package only
Relative price (single user)	PIC16 n/a PIC18 $495 PIC24 $895	PIC16 $995 PIC18 $995 PIC24 $1195	16F87X $50 PIC16 $150 PIC18 $200 PIC24 $250	PIC16+18 $249 PIC24 $249	PIC16 $99* PIC16+18 $180*
Demo version	Function-limited student edition	Time-limited evaluation version	Time- and memory-limited demo	None	None
Origin	US	US	US	EU	UK
*Approximate					

Microchip C18 and Hi-Tech C are designed primarily for professional use, as reflected in the relatively high price, but this is compensated for by the provision of feature-limited freeware versions. For any development engineer who will be using mainly PIC 18 or above parts, the C18 offers the advantage of extensive function libraries. Bear in mind though, a separate compiler, C30, is needed for PIC24 and dsPIC devices, although one can assume an easy progression route from C18. For those intending to use a wider range of MCU types, Hi-Tech might be preferred. Hi-Tech PICC Lite offers good functionality in a limited range of PIC 16 devices, including 16F877.

The educational compilers are designed primarily as components of training packages consisting of hardware, development system, compiler, tutorials, proprietary simulation software, and so on. These products should certainly be considered if a complete package is required, for example, by a college or university upgrading its resources. The Mikroelectronika packages are oriented more toward the hobby market, while the Matrix Multimedia product range is suitable for a wide range of education institutions, from schools to universities. The support materials provided with the Matrix Multimedia compiler are very closely tied to the training packages, so no separate compiler manual is provided, for example. For the hobbyist and independent learner, Mikro C is supported by a comprehensive and fully documented function library.

CCS C Programming Syntax Summary

Compiler Directives

```
#include source files
#use functions(parameters)
#define oldtext newtext
#device name
#list, #nolist
#asm, #endasm
#fuses options
#int_xxx
```

Include source code or header file
Include library functions
Replace label in source code with value
Identify MCU type
Turn on source code listing
Start/end of assembler block
Select MCU configuration fuse settings
Declare function as interrupt service routine

Program Blocks

```
main(condition) {statements}
while(condition) {statements}
do{statements} while(condition)
if(condition) {statements}
for(begin;end;next) {statements}
switch(x)..case n:
```

Main program block
Conditional loop
Conditional loop
Conditional sequence
Preset loop conditions
Multichoice selection

Punctuation

```
/* Comments */
statement; // Comment
{ statement; statement; }
statement;
funcname(arg1,arg2)
[n]
"text"
'y'
```

Star/slash enclose block comment
Double slash before line comment
Braces enclose program block
Semicolon = end of statement
Function arguments/parameters, comma separates
Array size, variable
ASCII function argument/include filename
ASCII value

Basic I/O Functions

output_X(n)	Output 8-bit code at Port X
output_high(PIN_Xn)	Set output bit high
output_low(PIN_Xn)	Set output bit low
input(PIN_Xn)	Get bit input
n=input_X()	Get byte input

Variable Types

Identifier	Type	Min	Max	Range
int1	1 bit	0	1	$1 = 2^0$
unsigned int8	8 bits	0	255	$256 = 2^8$
signed int8	8 bits	~127	~127	$255 = 2^8 - 1$
unsigned int16	16 bits	0	65,535	$65,536 = 2^{16}$
signed int16	16 bits	~32,767	~32,767	$65,535 = 2^{16} - 1$
unsigned int32	32 bits	0	4,294,967,295	$4,294,967,296 = 2^{32}$
signed int32	32 bits	~2,147,483,647	~2,147,483,647	$4,294,967,295 = 2^{32} - 1$
float	32 bits	$\sim 10^{-39}$	$\sim 10^{138}$	$\sim 10^{77}$

Relational Operators

Operation	Symbol	Example
Equal to	==	if(a == 0) b=b+5;
Not equal to	!=	if(a != 1) b=b+4;
Greater than	>	if(a > 2) b=b+3;
Less than	<	if(a < 3) b=b+2;
Greater than or equal to	>=	if(a >= 4) b=b+1;
Less than or equal to	<=	if(a <= 5) b=b+0;

Formatting Codes

Code	Displays
%d	Signed integer
%u	Unsigned integer
%Lu	Long unsigned integer (16 or 32 bits)
%Ls	Long signed integer (16 or 32 bits)
%g	Rounded decimal float (use decimal formatting)
%f	Truncated decimal float (use decimal formatting)
%e	Exponential form of float
%w	Unsigned integer with decimal point inserted (use decimal formatting)
%X	Hexadecimal
%LX	Long hex
%c	ASCII character corresponding to numerical value
%s	Character or string

Arithmetic and Logic Operators

1 Operand	Arithmetic, 2 Operands	Logic, 2 Operands
Assign value, =	Add, +	AND, &
Increment, ++	Subtract, -	OR, |
Decrement, --	Multiply, *	XOR, ^
Complement, ~	Divide, /	

CCS C Program Function Reference

This is a summary of the more commonly used functions available in CCS C Version 4 (January 2007). For more details on how to use the listed functions and others not included here, visit www.ccsinfo.com for a current manual download.

The following apply to all the following tables:

1. All functions require a header file, e.g., 16F877A.H.

2. The numerous CAN and USB functions are not included since these interfaces are not typically available in 16 series MCUs.

3. Alternative functions for the same operation:

```
putc() == putchar()
getc() == getch() == getchar()
output_bit(PIN_XX,1) == output_high(PIN_XX)
output_bit(PIN_XX,0) == output_low(PIN_XX)
get_timer0() == get_rtcc();
set_timer0(nnn) == set_rtcc(nnn);
pow() == pwr()
```

Table F.1: Port Input and Output
(Requires Chip Header File Only, e.g., 16F877A.H)

Function	Description	Example	Comment
WRITE BYTE	Write all bits with 8-bit integer	output_A(255);	A replaced by B, C, D, or E
SET BIT	Write output bit high using pin label	output_high (PIN_A0);	A0 replaced by A1, A2, . . ., A7, B0, . . ., B7, etc.
CLEAR BIT	Write output bit low using pin label	output_low (PIN_A0);	A0 replaced by A1, A2,. . ., A7, B0, . . ., B7, etc.
READ BYTE	Read input as 8-bit integer	abyte = input_A();	A replaced by B, C, D, or E
READ BIT	Read input bit using pin label	abit = input(PIN_A0);	A0 replaced by A1, A2, . . ., A7, B0, . . ., B7, etc.
READ DIRECTION	Check port data direction register	ddra = get_tris_a();	Any parallel port ddr code can be checked
CHECK BIT	Read input bit	abit = input_state (PIN_D0);	Gets I/O bit value
BIT TOGGLE	Toggle output bit	output_toggle (PIN_D0);	Invert the logic level at the specified pin
BIT OUTPUT	Change port bit to output	output_drive (PIN_D0);	Does not change the existing bit value
FLOAT OUTPUT	Set output pin to high impedance	output_float (PIN_D0);	Allows an external source to control the line
SET PULLUPS	Switch input pull-ups on or off	port_a_pullups (TRUE);	Input floats to high value, port A or B only
SET DIRECTION	Initialize port bits for input or output	set_tris_a (0x0F);	Explicitly sets up data direction register

Table F.2: Analog Inputs (Requires #DEVICE ADC = nn)

Function	Description	Example	Comment
SETUP	Initialize ADC	setup_adc(ADC_CLOCK_ INTERNAL);	All modes listed in device header file
PINS SETUP	Initialize ADC pins	setup_adc_ports (RA0_ANALOG);	All modes listed in device header file
CHANNEL SELECT	Select ADC input	set_adc_channel(0);	Channels 0–7 selected via multiplexer
READ	Read analog input	inval = read_adc();	8-bit read 0–255, 10-bit read 0–1024 (#device option)

Table F.3: Timers (Requires Chip Header File Only, e.g., 16F877A.H)

Function	Description	Example	Comment
TIMERX SETUP	Set up the timer mode	setup_timer0 (RTCC_INTERNAL | RTCC_DIV_8);	Clock source and prescale ratio
TIMERX READ	Read a timer register (8 or 16 bits)	count0 = get_timer0();	Timer numbers (0–5) valid as fitted
TIMERX WRITE	Preload a timer register (8 or 16 bits)	set_timer0(126);	Timer numbers (0–5) valid as fitted
TIMER CCP SETUP	Select PWM, Capture, or Compare mode	setup_ccp1 (ccp_pwm);	See CCS manual for CCP options
TIMER PWM DUTY	Set PWM duty cycle	set_pwm1_duty (512);	512 = mark count = 50%

Table F.4: RS232 Serial Port
(Requires #USE RS232, #USE DELAYS (Clock=nnnnnnnn))

Function	Description	Example	Comment
SET BAUD RATE	Set hardware RS232 port baud rate	`setup_uart(19200);`	Applies to hardware serial port only
SEND BYTE	Write a character to the default port	`putc(65)`	Writes ASCII data or control code to serial output
SEND SELECT	Write a character to selected port	`s = fputc("A",01);`	As preceding, but stream identifier given
PRINT SERIAL	Write a mixed message	`printf("Answer: %4.3d",n);`	Write fixed strings and formatted variable values
PRINT SELECT	Write string to selected serial port	`fprintf (01,"Message");`	As preceding, but stream identifier given
PRINT STRING	Print a string and write it to array	`sprintf (astr,"Ans=%d",n);`	Print and copy output to character array
RECEIVE BYTE	Read a character to an integer	`n = getc();`	Waits for ASCII code from serial input
RECEIVE STRING	Read an input string to character array	`gets(spoint);`	Reads characters into an array at address
RECEIVE SELECT	Read an input string to character array	`astring = fgets(spoint,01);`	As preceding, but string and stream identifier given
CHECK SERIAL	Check for serial input activity	`s = kbhit();`	Checks for serial input data but does not wait
PRINT ERROR	Write programmed error message	`assert(a<3);`	Generates an error message if condition is FALSE

Table F.5: SPI Serial Port (spi Can Be Replaced by spi2)

Function	Description	Example	Comment
SPI SETUP	Initialize SPI serial port	setup_spi (spi_master);	See CCS manual for full list of options
SPI READ	Receives data byte from SPI port	inbyte = spi_read();	Waits for 8-bit data to arrive
SPI WRITE	Sends data byte via SPI port	spi_write (outbyte);	Writes 8-bit data to SPI serial line
SPI TRANSFER	Send and receive via SPI	inbyte = spi_xfer (outbyte);	See CCS manual for variations
SPI RECEIVED	Check if SPI data received	done = spi_data_is_in();	Returns 0 for not done, 1 if done

Table F.6: I2C Serial Port
(#USE I2C() If Hardware Peripheral Fitted, #DEFINE for Software Interface)

Function	Description	Example	Comment
I^2C START	Issue start command in master mode	i2c_start();	Start a data transmission
I^2C WRITE	Send a single byte	i2c_write (outbyte);	Send a data byte
I^2C READ	Read a received byte	inbyte = i2c_read();	Read a data byte
I^2C STOP	Issue a stop command in master mode	i2c_stop();	Stop the data transmission
I^2C POLL	Check to see if byte received	sbit = i2c_poll();	Returns 1 if byte waiting

Table F.7: Parallel Slave Port

Function	Description	Example	Comment
PSP ENABLE	Enable or disable PSP	`setup_psp (PSP_ENABLED);`	`PSP_DISABLED` to switch off SET.
SET DIRECTION	Set the PSP data direction	`set_tris_e(0);`	For input arg. = 0xFF, or mixed mode
OUTPUT READY	Checks if output byte is ready to go	`pspo = psp_ output_full();`	Byte ready: `pspo = 1` To write the PSP: `PSP_DATA = outbyte;`
INPUT READY	Checks if input byte is ready to read	`pspi = psp_ input_full();`	Byte ready: `pspi = 1` To read the PSP: `inbyte = PSP_DATA;`
PSP OVERFLOW	Checks for data overwrite error	`pspv = psp_ overflow();`	Check to prevent loss of data due to external mistiming

Table F.8: LCD Control (Requires Chip Header File Only, e.g., 16F877A.H)

Function	Description	Example	Comment
LCD SETUP	Set up LCD internal control	`setup_lcd (LCD_MUX12,1);`	Number of control lines, clock prescale
LCD LOAD	Send display data block to LCD	`lcd_load (lcddata,0,16);`	Pointer, offset, number of bytes
LCD SYMBOL	Send segment bits	`lcd_symbol (lcddata,dig1)`	Specify segments individually

Table F.9: Register Manipulation

Function	Description	Example	Comment
REGISTER BIT SET	Set a selected bit	`bit_set(num,1);`	Sets bit b in integer num (8, 16, or 32 bits)
REGISTER BIT CLEAR	Clear a selected bit	`bit_clear(num,2);`	Clears bit b in integer num (8, 16, or 32 bits)
REGISTER BIT TEST	Test a selected bit	`flag = bit_test(num,4);`	Tests bit b in integer num (8, 16, or 32 bits)
REGISTER SWAP	Swap nibbles in a byte variable	`swap(abyte);`	Result not returned by function

Table F.10: Block Rotate

Function	Description	Example	Comment
BLOCK ROTATE LEFT	Rotates bits of structure left	`rotate_left (&lobyte,6);`	Address of low byte and number of bytes
BLOCK ROTATE RIGHT	Rotates bits of structure right	`rotate_right (&lobyte,10);`	Address of low byte and number of bytes
BLOCK SHIFT LEFT	Shift bit left into low bit of structure	`shift_left (&lobyte,4,1);`	Address of low byte, number of bytes, bit in
BLOCK SHIFT RIGHT	Shift bit right into high bit of structure	`shift_left (&lobyte,4,1);`	Address of low byte, number of bytes, bit in

Table F.11: Math Functions (`#INCLUDE MATH.H`)

Function	Description	Example	Comment
ABSOLUTE VALUE	Absolute value of integer	`abres = abs(x);`	Returns unsigned positive value of signed integer
LONG ABSOLUTE	Absolute value of long integer	`longres = labs(x);`	Returns unsigned positive value of 16-bit integer
FLOAT ABSOLUTE	Absolute value of float	`flores = fabs(x);`	Returns unsigned positive value of signed float
FLOAT CEILING	Round a float up to integer	`roundup = ceil(afloat);`	Returns integer from float
FLOAT FLOOR	Round a float down to integer	`roundown = floor(afloat);`	Returns integer from float
INTEGER DIVIDE	Integer divide	`divres = div(numer,denom);`	Returns a structure of quotient and remainder
LONG DIVIDE	Long integer divide	`lonres = ldiv(lnumer,ldenom);`	Returns a structure of quotient and remainder
EXPONENTIAL	Exponential function	`expres = exp(x);`	Returns exp where x is a float

(continued)

Table F.11: (continued)

Function	Description	Example	Comment
LOG BASE 10	Logarithm base-10 function	`logres == log10(x);`	Returns `log10(x)` where x is a float
LOG BASE E	Logarithm base-e function	`lnres = log(x);`	Returns `ln(x)` where x is a float
DIVISION MODULUS	Modulus (remainder) of division	`modres = fmod(numer,denom);`	Returns remainder of float division
FRACTION MODULUS	Break up float into integer and fraction	`modfres = modf(afloat,&whole);`	Returns fractional part, stores integer
FRACTION EXPAND	Break up float into integer and fraction	`fexres = frexp(afloat,&whole);`	Returns fractional part
BINARY EXPAND	Multiply a float by integral power of 2	`lexres = ldexp(afloat,sint);`	Returns a float, `sint` is a signed integer
RAISE TO POWER	Raise float to a power	`powres = pow(afloat,apower);`	Returns a float raised to a power
SQUARE ROOT	Calculate the square root of a float	`sqrres = sqrt(afloat);`	Returns positive root
RANDOM NUMBER	Generates a pseudorandom number	`any1 = rand();`	Returns a random integer from sequence
RANDOM SEED	Start value for the "random" sequence	`srand(seed);`	seed is a new start point in the sequence

Table F.12: Trigonometric Functions (`#INCLUDE MATH.H`)

Function	Description	Example	Comment
SIN	Sine function	num1 = sin(a);	Returns sine of angle a given in radians
COS	Cosine function	num2 = cos(a);	Returns cosine of angle a given in radians
TAN	Tangent function	num3 = tan(a);	Returns tangent of angle a given in radians
ASIN	Arc sine function	ang1 = asin(n);	Returns the angle in radians whose sine is float n
ACOS	Arc cosine function	ang2 = acos(n);	Returns the angle in radians whose cosine is float n
ATAN	Arc tangent function	ang3 = atan(n);	Returns the angle in radians whose tangent is float n
SINH	Hyperbolic sine function	hyp1 = sinh(x);	Returns hyperbolic sine of float x
COSH	Hyperbolic cosine function	hyp2 = cosh(x);	Returns hyperbolic cosine of float x
TANH	Hyperbolic tangent function	hyp3 = tanh(x);	Returns hyperbolic tangent of float x

Table F.13: Make Integers

Function	Description	Example	Comment
MAKE BYTE	Extract a byte from long integer	mybyte = make8(num,3);	Extracts byte from 16- or 32-bit integer
MAKE WORD	Make a 16-bit integer	myword = make16(byte1, byte0);	Combine separate bytes into one integer
MAKE LONG	Make a 32-bit integer	mylong = make32 (byte3,byte2, byte1,byte0);	Combine 4 bytes or two 16-bit integers

Table F.14: Type Conversions (#INCLUDE STDLIB.H)

Function	Description	Example	Comment
ASCII TO FLOAT	ASCII to float conversion	num0 = atof(decstring);	Converts a decimal number as string into float
ASCII TO INTEGER	ASCII to 8-bit integer conversion	num1 = atoi(intstring1);	Converts an integer given as string into an 8-bit integer
ASCII TO LONG	ASCII to 16-bit integer conversion	num2 = atol(intstring2);	Converts an integer given as string into a 6-bit integer
ASCII TO 32 BIT	ASCII to 32-bit integer conversion	num3 = atoi32(intstring3);	Converts an integer given as string into a 32-bit integer

Table F.15: Character Test (#INCLUDE CTYPE.H)

Function	Description	Example	Comment
ALPHANUMERIC?	Test for alphanumeric character	test = isalnum(acode);	Returns 1 if character code is in ranges 0–9, A–Z, a–z
NUMBER DIGIT?	Test for numerical digit character	test = isdigit(acode);	Returns 1 if character code is in range 0–9
LOWER CASE?	Test for lower case alphanumeric	test = islower(acode);	Returns 1 if character code is in range a–z
SPACE?	Test for space character	test = isspace(acode);	Returns 1 if character code is a space
UPPER CASE?	Test for upper case alphanumeric	test = isupper(acode);	Returns 1 if character code is in ranges A–Z
HEX DIGIT?	Test for hexadecimal digit	test = isxdigit(acode);	Returns 1 if character code is in ranges 0–9, A–F, a–f
CONTROL?	Test for control character	test = iscntrl(acode);	Returns 1 if character code is control code (00 – 1F)
GRAPHIC?	Test for printable character	test = isgraph(acode);	Returns 1 if character code is graphical (21 – 7E)
PRINTABLE?	Test for printable or space character	test = isprint(acode);	Returns 1 if character code is printable (20 – 7E)
PUNCTUATION?	Test for punctuation character	test = ispunct(acode);	Returns 1 if character code is a punctuation code

Table F.16: Search and Sort (#INCLUDE STDLIB.H)

Function	Description	Example	Comment
BINARY SEARCH	Search for given value in a data array	bsearch (k,a1,n,w,compit)	Find value k in array a1 of n elements of width w
QUICK SORT	Sort an array into ascending order	qsort (a1,n,w,sort1)	Sort array a1 of n elements of width w using function sortit

Table F.17: Processor Controls
(Requires Chip Header File Only, e.g., 16F877A.H)

Function	Description	Example	Comment
GET ENVIRONMENT	Gets information about the MCU	chip = getenv(device);	Peripheral hardware, memory, configuration, etc.
GOTO ADDRESS	Jump to program memory location	goto_ address(0x1FF0);	Jump in ROM, use with caution
LABEL ADDRESS	Check address of program label	labloc = label_address (start);	Labels should be used only in exceptional cases
RESET CPU	Restarts the program from 0	reset_cpu();	No return
RESTART CAUSE	Returns cause of last reset	message = restart_cause();	Messages defined in MCU header file
RESTART WATCHDOG	Clear watchdog timer	restart_wdt();	Periodical operation to prevent MCU watchdog reset
SETUP OSCILLATOR	Select internal clock mode	setup_ oscillator();	MCUs with internal clock
SLEEP	Stops program and waits for reset	sleep();	Wake up on specific events

Table F.18: Interrupts
(Requires Chip Header File, e.g., 16F877A.H & #INT_XXXX)

Function	Description	Example	Comment
INTERRUPT DISABLE	Disables peripheral interrupt	`disable_interrupts (int_timer0);`	Interrupt labels defined in device header file
INTERRUPT ENABLE	Enables peripheral interrupt	`enable_interrpts (int_timer0);`	Interrupt labels defined in device header file
INTERRUPT CLEAR	Clears peripheral interrupt	`clear_interrupt (int_timer0);`	Interrupt labels defined in device header file
INTERRUPT ACTIVE	Checks if interrupt flag is set	`interrupt_active (int_timer0);`	Interrupt labels defined in device header file
INTERRUPT EDGE	Selects interrupt trigger edge	`ext_int_edge (H_TO_L);`	Rising (L_TO_H) or falling (H_TO_L) edge
INTERRUPT JUMP	Jump to address of ISR	`jump_to_isr (isr_loc);`	Use to service multiple interrupts

Table F.19: Memory Read and Write

Function	Description	Example	Comment
READ RAM BANK	Read a RAM location directly	`abyte = read_bank(3,0x20);`	Alternative variable access
WRITE RAM BANK	Write a byte into user RAM	`write_bank (3,0x20,0xFF);`	Write to bank 3, address 0x20, data 0xFF
READ DATA EEPROM	Read an EEPROM location	`abyte = read_eeprom(0x00);`	Get byte at given address
WRITE DATA EEPROM	Write a byte into EEPROM	`write_eeprom (0x1F,0x9A);`	Write to nonvolatile memory address, data
READ PROGRAM ROM	Read code from program ROM	`read_program_memory (0x100,copy,4);`	Get block from program address, copy in RAM

Table F.20: Memory Allocation (#INCLUDE STDLIBM.H)

Function	Description	Example	Comment
MEMORY BLOCK ALLOCATE	Reserves a block of memory	ap1 = calloc(25,4);	Allocated block = 25×5 bytes
MEMORY BLOCK DEALLOCATE	Releases a memory block	free(ap1);	Previously allocated at address pointer ap1
MEMORY BYTES ALLOCATE	Reserves a number of bytes	ap1 = malloc(14);	Allocated block = 14 bytes
MEMORY BLOCK COPY	Copy a given number of bytes	memcpy (ap1,ap2,n);	Copy n bytes from ap1 to ap2
MEMORY BLOCK MOVE	Move a given number of bytes	memmove (ap1,ap2,n);	Move n bytes from ap1 to ap2
MEMORY BLOCK SET	Initialize locations with a given value	memset (ap1,val1, numofb);	Loads integer val1 into numof locations from ap1

Table F.21: Special Setup
(Requires Chip Header File Only, e.g., 16F877A.H)

Function	Description	Example	Comment
SETUP WATCHDOG TIMER	Initialize watchdog time-out	setup_wdt (wdt_1152ms);	Time-out options from 18 ms to 2.304 sec
RESET WATCHDOG TIMER	Clear watchdog timer within the program loop	restart_wdt();	Watchdog timer is normally reset before time-out
SETUP COMPARATORS	Connection of analog comparators	setup_ comparator (A0_A3_A1_A2);	Selected MCUs only
VOLTAGE REFERENCE	Specify the comparator ref. voltage	setup_vref (vref_low|10);	Options in device header file
SETUP OPAMP	Enable built in op-amp where fitted	setup_ opamp1(1);	Selected MCUs only
SETUP SLEEP	Sets sleep delay time	sleep_ulpwl (time_in_us);	Selected MCUs only
LOW VOLTS DETECT	Triggers interrupt if supply low	setup_low_ volt_detect (lvd_33);	Selected MCUs only

Answers

Assessment 1

1. Musical birthday card, electronic price tag, sound system, television, automobile, robot.

2. Input, ROM, CPU, RAM, output.

3. Flash ROM is non-volatile but reprogrammable, so the program can be changed or the chip reused. Program testing and modification is easier and development time is reduced compared with alternative types of program memory.

4. Number of I/O pins, program memory size, RAM size, EEPROM size, maximum clock speed, range of interfaces, development system, cost, availability.

5. The program is stored as machine code instructions, executed in sequence. The instruction register holds the current instruction and the program counter holds its address. The file registers store the program data and the working register the data being operated on.

6. 02 = Program Counter Low Byte.
   03 = Status Register.
   09 = Port E Data Register.
   89 = Port E Data Direction Register.
   20 = General Purpose Register 1.

7. RC = clock uses resistor/capacitor circuit to control clock frequency.
   XT = clock uses crystal circuit to control clock frequency.
   WDT = watchdog timer provides automatic reset if program hangs.
   PUT = power-up timer delays the program start until the MCU is ready.
   NOWRT = prevents writes to program memory areas.

8. Tristate gate = data switching circuit allows data through only when enabled; otherwise, output is high impedance.
Current driver = provides extra current on a loaded data line.
Data direction latch = stores the bit that sets the port bit as input or output.
Input data latch = stores the incoming bit when the port line is set to input.
Output data latch = stores the outgoing bit when the port line is set to output.

9. 20-MHz clock → 5-MHz instruction clock → 200-ns period.
10 ms = 10,000,000 ns.
Timer count required = 10,000,000/200 = 50,000 instruction clock cycles.
Maximum count of 16-bit timer = 65,536.
Preload value = 65,536 − 50,000 = 15,536.

10. Resolution = 2048/256 = 8 mV per bit.
Output = (1000/2048) × 256 = 125 = 0 + 64 + 32 + 16 + 8 + 4 + 0 + 1
= $0111\ 1101_2$.

11. The timer interrupt is set up at the beginning of the program. The timer is started at some point in the program and runs concurrently with program execution. When a time-out occurs, the program is suspended and the interrupt service routine carried out. The program is then resumed at the original point. Interrupts allow the timer to independently generate an accurate interval between the timer start and interrupt request.

12. See the figure.

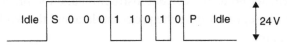

13. RS232 is asynchronous, in that it has no separate clock signal. Instead, the reception is resynchronized by each start bit, and reception is timed by a local clock. SPI has a separate clock (used to strobe each bit into the receiver, generated from the master MCU clock) and is therefore classed as a synchronous system.

14. SPI needs a hardware chip select signal connected to each slave, which the master takes low to enable one slave receiver at a time. I^2C transmits the target address on the data line; the slave must check all transmissions and pick up the data that follow its own address.

15. I^2C has to send addressing and control information as well as the data on the data line, while SPI has hardware slave selection.

16. RS232 = 9600 baud ≈ 10 k bits/sec ≈ 1 k bytes/sec ≈ 1000 characters/sec →
    Page time ≈ 1 sec.
    SPI = 5-MHz clock → 0.2 μs/bit → 2 μs/character (some loading delay) →
    Page time ≈ 2 ms.

17. `C` = source code entered via a text editor.
    `HEX` = hexadecimal code (machine code) program.
    `COF` = downloading file that contains the hex code plus debugging information.
    `LST` = list file, a text file containing source code, hex code, comments, etc.
    `ERR` = error file that lists the error messages generated by the compiler.

18. V_{ss} = 0 V, V_{dd} = +5 V = supply connections.
    V_{pp} = programming voltage (+14 V); `!MCLR` = Master Clear resets MCU.
    `PGD` = program data download; `PGC` = programming clock signal.

19. Project file = shows the files used to make the project.
    Source code = edits window for entering program.
    Disassembler list file = shows the assembler code generated from the C source
    code.
    Output message = shows the compiler status and errors.
    Watch = variable values monitored during program execution.

20. Host PC, MPLAB development system, C compiler, programming
    module + connectors, target system with PIC MCU.

Assessment 2

1. Include header file using `#directive`.
   Main program statement block enclosed in braces.
   I/O functions/sec within main.

2. Create MPLAB project.
   Edit program using correct syntax.
   Build program and correct syntax errors.
   Test program in simulator and debug.
   Optional—test in cosimulation mode.

3.
```
output_C(64);
output_high(PIN_C6);
```

4. The WHILE loop tests the control condition before the loop statements are executed. The DO..WHILE tests after they have been executed at least once. The FOR loop executes a loop a fixed number of times.

5. Port D bits initially go on for 1 sec. If the switch is active, the high 4 bits then go off, and the program waits until the switch goes inactive, at which point all the outputs go off. If the switch is inactive, all the LEDs go off after 1 sec.

6. (a) 255 (b) 32,767 (c) $(2 - 1/2^{23}) \times 2^{128} = 6.8 \times 10^{38}$

7. (a) 8-bit precision $= 1/2^8 \times 100\% = 0.39\%$.
   (b) 32-bit FP precision $= 1/2^{23} \times 100\% = 0.000012\%$.

8. Mantissa $= 011 \rightarrow \frac{1}{4} + \frac{1}{8} = 0.25 + 0.125 = 0.375 \rightarrow 1.375$.
   Exponent $= 1000\ 0010 = 130 \rightarrow 130 - 127 = +3 \rightarrow 2^3 = 8$.
   Sign $= 0 \rightarrow$ positive.
   Number $= 8 \times 1.375 = 11.000000$.

9. ```
   a = n + 0x30;
   putc(a);
   ```

10. $n = 5 = 0101_2$, $m = 7 = 0111_2$.
    (a) 6, 0110 (b) 8, 1000 (c) 5, 0101 (d) 7, 0111 (e) 2, 0010

11. Continue means restart a loop, Break means quit a loop, Goto means jump to a label unconditionally.

12. ```
    switch(x)
    {   case 1: fun1();
        break;
        case 2: fun2();
        break;
        case 3: fun3();
        break;
    }
    ```

13. Local variables are allocated memory only when a function is called and are discarded when the function has finished. The memory can then be used for other purposes, saving on overall memory requirements.

14. Functions are self-contained blocks that implement a clearly defined set of operations, receiving data for processing and returning results to the calling routine. A structured program is a nested or hierarchical set of functions that

is easy to understand and modify. Reusable function libraries can be created, which save on programming time. Compiler packages provide function libraries for the most common operations.

15. `int` = variable type returned.
    `out` = name of the function.
    `int16 t` = variable and type received.
    `int16 n` = local variable declaration.
    `outbyte` = value returned from function.

16. The RS232 signal has a start bit, 8 data bits, and a stop bit. The edge of the start bit triggers the LCD receiver shift register to sample the line in the middle of each data bit. This is stored as an ASCII character and displayed. Control codes for the LCD are preceded by the code 254.

17. See the figure.

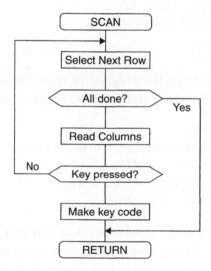

18. The function prints formatted output. This means that any variable output has an associated formatting code, such as %d, which determines how the value is interpreted. The main options are signed integer, floating point decimal, or ASCII character. The variable `anum` in this case is an array variable, the element being output is numbered `n`.

19. Ampersand (&) is the address_of operator, which causes the memory address of the named variable to be returned. The pointer (*) is the contents of operator, which returns the value of the contents of the location corresponding to the variable value.

20. `#include` means copy another source code file into the user source code, `#define` instructs the compiler to replace the given text with the given value, `#use` means include a library function, `#device` defines the target MCU and optionally an operating mode, `#asm` indicates the start of an assembly language sequence,

Assessment 3

1. `setup_adc_ports(AN0);`
   Reference = 5V, resolution = 5/1024 = 4.88 mV/bit.

2. Resolution = 4.096/1024 = 4.00 mV/bit, conversion factor = 0.004.
   (a) `float volts,input;`
   (b) `volts=input*0.004;`

3. `enable_interrupts(int_AD);`
   `enable_interrupts(global);`
   `#int_AD`
   `void isrADC(){}`

4. Using the ADC interrupt, the program is more efficient because time is not wasted in polling the ADC, and the ADC result can be processed as soon as it is available.

5. 16-bit maximum count = 65,536, remaining count = 65,536 − 15,536 = 5,000.
   Instruction clock = 8/4 = 2 MHz.
   Clock period after prescale = 16/2 = 8 μs.
   Timer period = 5000 × 8 = 40 ms.

6. The Capture mode uses an input bit change to trigger the capture of the current timer reading, transferring it into the preload registers for processing. This mode can be used for input signal period measurement. The Compare mode needs the preload registers to be loaded with a value with which the current timer value is continuously compared. An interrupt flag is set and an output toggled when they match. This mode can be used to generate an output of a given period.

7. See the figure.

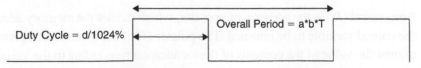

Duty Cycle = d/1024%

Overall Period = a*b*T

8. Output period $= 1000\,\mu s = 1000$ clocks $= 250 \times 4 =$ timer count $\times$ prescale $\rightarrow$
   `setup_timer_2(4,250,1)`.
   10% duty cycle $= 102/1023 \rightarrow$ `set_PWM1_duty(102)`.

9. The standard serial LCD is designed to receive 8-bit ASCII codes in RS232 format. High speed is not required, because only a limited amount of data is sent as the display is updated. The longer-link distance possible with RS232 may be useful if the display is mounted away from the MCU board.

10. `0x41` is the ASCII code for character 'A'. In the `printf()` statement, it is output and displayed as a decimal 65 because the formatting code is `%d`. The `putc()` function outputs the ASCII code and displays the character 'A'.

11. The UART data transfer takes about 1 ms, during which time the MCU could be working on another task. MCU utilization can be increased by using interrupts, which can be set up to fire when the serial port has finished sending (`int_tbe`) or receiving (`int_rda`) a byte. The interrupt service routines contain the code to write the next byte or read the next byte. On return from interrupt, a foreground task continues, which is interrupted again only when the UART is ready for the next byte transfer.

12. Each slave sender needs a slave select line connected to the master MCU, not to ground. The master program contains bit switching statements to enable the select line of a slave MCU programmed to transmit.

13. ```
i2c_start();
i2c_write(0xA0);
i2c_write(0x01);
i2c_write(0xFF);
i2c_write(0xAA);
i2c_stop();
```

14. See the figure.

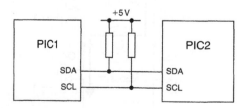

15. Set up the PSP interrupt in the slave PIC.
    Select the slave PIC by taking !CS low.
```

Present the data to the Port D data pins.
Take !WR low to latch in the data.
Interrupt INT_PSP generated to read the port data.

16. The minimum number of wires is used by I^{2}C, but the rate of transfer is reduced compared with SPI because control and address bytes have to be sent before the data are returned.

17. (a) PSP (b) SPI (c) UART (d) I^{2}C

18. EEPROM is nonvolatile data storage, which allows data to be stored while the power is off. It can therefore store security codes and limited amounts of other key data long term. It is limited in size, so an external serial EEPROM can be used to expand it.

19. The output speed is critical in this application, because the waveforms are generated by outputting a table of values to the DAC as fast as possible. To minimize the output loop time, interrupts are used instead of polling the switches. The output frequency is thereby maximized.

20. An output bit can be toggled using an assembler sequence to minimize the loop time, as shown in Section 2.8. In this circuit, the output port needs to be switched between `0x00` and `0xFF` using `output_portD(n)` within a minimal loop to generate a fast square wave.

Assessment 4

1. See the figure.

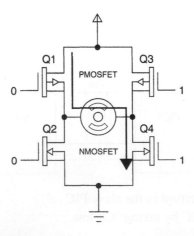

2. Speed = 6 steps/sec = 6 × 7.5 deg/sec = 45 deg/sec = 45/360 rev/sec →
   60/8 = 7.5 rpm.

3. Linear characteristic: Output voltage, V_t = mt + c; t = temperature; Gradient,
   m = 10 mV/°C.
   At 0°C, sensor voltage, V_t = 500 mV, so 500 = c. Hence, V_t = 10t + 500 mV.

4. The parallel display uses more MCU output pins, drive requirements are more
   complex (segment encoding required), and it shows only 3.5 numerical digits,
   while the serial LCD is `16x2` alphanumeric.

5. `lcd_symbol(DigMap[8],DIG1);`
   The first argument of the function is an array variable that contains the seven-
   segment code for the number 8, and the second identifies the seven display
   memory bits for the segments of the digit.

6. The DC motor needs position feedback to achieve a set position or speed. A
   slotted wheel and optical sensor produce pulses as the shaft turns, allowing the
   MCU to count the revs completed in unit time.

7. The stepper motor has multiple coils, which are energized in sequence to turn
   the shaft, so it can be turned through a set number of steps with no feedback
   required. The stepper motor on the mechatronics board has two sets of
   windings, two wires each, which are connected to the four drive outputs.

8. Connect the motor sensor to Timer1 input and configure the timer to measure
   the pulse period. The Capture mode of operation allows the timer count to be
   captured when the sensor input changes. The MCU program can convert the
   pulse period into revs/sec.

9. 1 step = 7.5°, 1 rev = 360/7.5 = 48 steps.
   Time per step = 1/48 = 20.8 ms ≈ 21 ms.

10. The temperature sensor gives an output of 10 mV/°C, with an offset of 500 mV,
    so the temperature can be calculated at any value in that range. The light sensor
    output cannot be quantified in the same way, because it is not linear and the
    absolute level is therefore more difficult to calculate.

11. V_t = sensitivity × temp + offset = (10 × 25) + 500 = 750 mV.
    ADC output scaling = 2048/1024 = 2 mV/bit.
    ADC output value = V_t/scaling = 750/2 = 375.

12. Sink = `Pg.Ng` = `M.N.F`.
    Source = `!Pg.!Ng` = `(P.M.!N.F)+(P.!M.F)` = `P.F.((M.!N)+!M)`.

13. P1 and N2, P2 and N1, M1 and M2. The current flows diagonally across the bridge, so P1 and N2 are on together for forward current and P2 and N1 for reverse. M1 switches on and off N2 and M2 switches N2 for PWM control.

14. With the inputs linked for full bridge operation, P1 and P2 operate Drives 1 and 2, respectively, which are connected to stepper motor Coil 1, brown and orange wires. PWM1 is connected to CCP1 output. P3 and P4 operate Coil 2, red and yellow; and PWM3 is connected to CCP2. Sequence: Drive 1, 4, 2, 3.

15. It is voltage operated with a high input impedance, so it is simple to interface and can be driven directly from a logic output. The output 'on' resistance is low, and the 'off' resistance is high.

16. Gain of amp = 10.
    Sensing resistor = $0.1\,\Omega$.
    Test resistor = $3.3 + 0.5 = 3.8\,\Omega$.
    Total resistance = $3.8 + 0.1 = 3.9\,\Omega$.
    Amp input voltage = $(0.1/3.9) \times 5 = 0.13\,V$.
    Amp output voltage = $0.13 \times 10 = 1.3\,V$.

17. The latch consists of cross-coupled NOR gates, such that only one output can be high at a time. The drives are disabled when the fault output is low and the LED output is high. The comparator output goes high when an overcurrent is detected, forcing the fault output low and switching on the LED. This state is held until the Reset button forces the LED output low and the fault output high, resetting the latch.

18. The MOSFET is switched by applying 5 V between the gate and source, with the load connected to the drain. The NMOSFET has its source connected to 0 V and is switched on with 5 V at the gate; the PMOSFET has its source connected to +5 V and is switched on with 0 V at its gate. This provides symmetrical drive components in the half bridge.

19. Connect the motor between Drives 1 and 2. Enable drive at P1 from MCU RD7, and control N2 from MCU CCP2(RD2). PWM output is generated from the CCP2 module, which controls the speed of the motor.

20. Output sequence at Port D: 0x80, 0x10, 0x40, 0x20.
    PWM inputs not connected = 1 (enabled).
    Outputs high: RD7(P1 + N2), RD4(P4 + N3), RD6(P2 + N1), RD5(P3 + N4).
    Drive sequence: Winding1 forward(Drive1 → Drive2).
                     Winding2 reverse(Drive4 → Drive3).
                     Winding1 reverse(Drive2 → Drive1).
                     Winding2 forward(Drive3 → Drive4).

Assessment 5

1. Hysteresis means that the switching level of the input depends on the polarity
   of the input change. This helps overcome noise on the input, which would cause
   unreliable switching, by implementing an upper and lower switching levels.

2. `set_adc_channel(0);`
   `numin = read_adc();`
   In the read statement, the input value returned *by* the function has to be
   assigned to another variable for processing. In the channel select statement, the
   channel number is passed *to* the function as the function argument.

3. See the figure. (10 points)

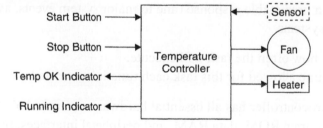

4. `TEMPCON`

   ```
   Initialize
     MCU, ADC, Functions
   Wait for 'Start'
   Switch on 'Running'

   Loop
     Read temperature
     If too low
       Switch on Heater
     If too high
       Switch on Fan
   ```

```
If OK
   Switch on 'TempOK'
Always //
```
(10 points)

5. A data logger often needs to record analog input values from sensors. Flash ROM is nonvolatile so data are retained during power off, and the serial interface uses only two pins on the MCU. A serial link is needed to upload the acquired data to a host system.

6. In a polled system, the time between input samples may vary if the processing time changes between samples. A timer interrupt forces the execution of an ISR containing the input sampling event at fixed intervals.

7. In a system with multiple interrupts, each is assigned a numerical priority in relation to the others, such that a high-priority ISR is not interrupted by a lower-priority one, but a low-priority interrupt may be interrupted by a high-priority task.

8. The PC operating system is a priority-interrupt driven, multitasking OS optimized for file processing, so that the time response of the system to real-time events is not predictable. The real-time operating system is designed to provide a predictable response time to major system events, as required in control systems.

9. `rate` = how often the task will execute.
   `max` = time allowed for this task each time it is executed.

10. The microcontroller has all essential hardware resources built into one chip: CPU, program ROM, data RAM, and peripheral interfaces. In a conventional microprocessor system, these are provided as separate chips so that the system can be tailored to the application.

11. The system on a chip allows the microcontroller hardware to be configured for a specific application then manufactured on one chip, giving the benefits of both the conventional microprocessor system and the microcontroller.

12. Familiarity, cost, complexity, range, development system, availability, features.

13. Sufficient I/O pins, peripheral support, program memory size, data memory size, speed, power consumption.

14. The prototype costs are mainly hardware and software design time. As more units are produced, the development costs are shared, so that the cost per unit falls with the volume of production (see the figure).

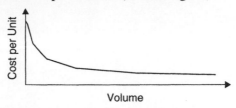

15. The serial alphanumeric LCD needs only a single MCU pin and can display several lines of numbers and characters. The 3.5-digit LCD is cheaper, the digit display is larger, and access is faster.

16. The size of the system and number of components largely determine the power consumption, plus the current drawn by the MCU increases with the clock speed. The component data sheets need to be consulted to predict power consumption, as this is not generally modeled in simulation systems. A prototype must be built to confirm the power supply specification.

17. C is a higher-level language than assembler, so it is easier to learn and use, as the meaning of the program statements is more obvious. The same standard C syntax is used for all processors, with the compiler converting the source code into the MCU-specific assembly language. This means that it is universal and, to some extent, portable between systems. The basic programming techniques are applicable to all embedded systems, with the main variation being in the I/O function syntax.   (10 points)

Index

Printed and bound by CPI Group (UK) Ltd, Croydon, CR0 4YY

03/10/2024

01040336-0006